From the Sun

Auroras, Magnetic Storms, Solar Flares, Cosmic Rays

Steven T. Suess
Bruce T. Tsurutani
Editors

American Geophysical Union
Washington, DC

Published under the aegis of the AGU Books Board

From the Sun: Auroras, Magnetic Storms, Solar Flares, Cosmic Rays
Steven T. Suess and Bruce T. Tsurutani, Editors

Library of Congress Cataloging-in-Publication Data
From the sun: auroras, magnetic storms, solar flares, cosmic rays /
    Steven T. Suess, Bruce T. Tsurutani, editors.
        p. cm.
        Includes bibliographical references.
        ISBN 0-87590-292-8
        1. Sun. 2. Solar wind. I. Suess, Steven T. II. Tsurutani,
Bruce T.
QB521.6.F76 1998   98-46324
523.7--dc21 CIP

Cover: (Front) Bright streamers and a corkscrew-shaped coronal mass ejection appear in this image of the normal K- or electron corona. The twisted mass of ionized gas, expelled from the lower atmosphere, and contorted by the magnetic fields that hold it together, is seen stretching across the field of view out to more than three million kilometers above the Sun's visible surface. The blue image superimposed on the center of the picture shows the hot ionized gas in the low solar corona at nearly the same time as the larger image. (Back) This image of the Earth's northern auroral zone, taken aboard the POLAR spacecraft, shows the onset of a geomagnetic substorm.

The chapters herein were modified from articles originally published in *Eos Transactions* of the American Geophysical Union. The publication dates for the original articles are: "Aurora," May 12, 1992; "The Earth's Magnetosphere," Dec. 19, 1995; "Radiation Belts," August 20, 1991; "Plasma Waves and Instabilities," Dec. 8, 1992; "The Ionosphere and Upper Atmosphere," March 12, 1996; "Red Sprites and Blue Jets: Transient Electrical Effects of Thunderstorms on the Middle and Upper Atmospheres," Jan. 2, 1996; "Magnetic Storms," February 1, 1994; "The Human Impact of Solar Flares and Magnetic Storms," Feb. 18, 1992; "The Solar Wind," May 18, 1993; "Solar Flares," Nov. 23, 1993; "Solar Flare Particles," Oct. 4, 1994; "Solar Irradiance Variations and Climate," Aug. 16, 1994; "The Solar Dynamo," Nov. 22, 1994; "Cosmic Rays," March 7, 1995; "Anomalous Cosmic Rays: Interstellar Interlopers in the Heliosphere and Magnetosphere," April 19, 1994; and "The Outer Heliosphere," Dec. 13, 1994.

Printed in the United States of America

Contents

Preface

A number of years ago, it became apparent that there was no single book or pamphlet that described space plasma physics at a level that a layperson with a high school physics and chemistry background could understand. There are several textbooks available, but these are targeted toward college undergraduate or graduate students and usually require knowledge of advanced mathematics. So, with the support of the American Geophysical Union, we have asked some of the world's experts in this field to donate their time and to write short chapters for this book. They have attempted to write brief summaries of their area of specialty with little or no mathematics (for the most part, they succeeded!), at a level for a high school student interested in science. The topics were chosen to address obvious questions that the person in the street might have (what is a Radiation Belt?, What are Auroras?, What Causes Solar Flares?, How Does the Sun Work?, What are Magnetic Storms?, What is the Human Impact of Solar Flares and Magnetic Storms?, etc.) and at the same time cover the fields that NASA, the National Science Foundation, the National Oceanic and Atmospheric Administration, the European Space Agency, and other agencies fund for research. The success of this book is due to the efforts of all the individual authors. We, as editors, hope the selection of topics covers the field of space physics adequately.

A glossary has been appended to the original articles at the behest of the authors and some of the readers of *Eos*. There are many unusual and unfamiliar terms in space physics and the glossary has already been a useful addition for us. We expect this will be one of the more useful parts of this book for the general reader.

This book should also provide quick reading and useful reference for the general AGU (nonspace physics) membership, space physics graduate students wishing to survey different fields of research, and space physics researchers interested in finding out what the "others" are doing. It was extremely beneficial to us to read all of the articles and to comment on what concepts/phenomena were easy/hard to follow and how one might possibly do it better. Finally, we hope that this book will be useful to give to (non-scientific) friends and relatives who ask what we space scientists do. Well-placed bookmarkers at the appropriate chapters will allow them to obtain quick overviews and hopefully an appreciation of our research contributions.

Steven T. Suess and Bruce T. Tsurutani

Aurora

Syun-Ichi Akasofu

The aurora is the luminous emission of atoms and molecules in the polar upper atmosphere that appears as permanent, ring-shaped belts called the auroral oval around the north and south geomagnetic poles (see Figures 1a and 1b). It is associated with a global electrical discharge process that requires about 1 million MW or more. In fact, electric currents of a few million amperes flow along the auroral oval (Figure 2). Except for specialized books for professionals in this field, however, most books describe the aurora as a result of the direct entry of energetic particles from the Sun toward the polar region. This view is too simplistic in terms of the progress we have made in this field since the advent of the satellite age. Indeed, we have begun to understand the generator process that powers the auroral discharge. In this article, the auroral process is described in terms of this latest knowledge.

The Sun emits an enormous amount of energy in the form of charged particles in addition to the 6000 °K blackbody radiation that is crucial for life on Earth [*Foukal*, this vol.]. This particular flow of charged particles is called solar wind [*Goldstein*, this vol.]. It originates in the corona, the uppermost part of the solar atmosphere, which has a temperature of more than 1,000,000°. Because of its high temperature, the coronal gas is ionized. An ionized gas is called a plasma. Therefore, the solar wind is a supersonic plasma flow that consists mainly of protons and electrons.

At the distance of the Earth, the speed of the solar wind is usually about 400–500 km per second. At the present time, there is no accepted theory on the generation of the solar wind. The Voyager spacecraft is still detecting the solar wind at the distance of Neptune and beyond [*Axford and Suess*, this vol.].

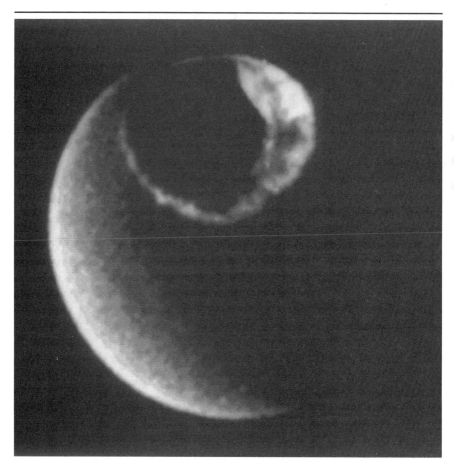

Figure 1a. The auroral image obtained by the Dynamic Explorer satellite from above the north pole region (courtesy of J. Craven and L. Frank, University of Iowa).

The Earth is immersed in the hot solar wind. Here, the Earth's magnetic field plays an important role. It was William Gilbert who discovered in 1600 that the Earth itself is a gigantic magnet, although geophysicists are still struggling to understand how the Earth is magnetized. It was thought once that there was a bar magnet in the Earth. This idea was soon abandoned after researchers found that the temperature of the interior of the Earth is well beyond the melting point of iron. Now it is generally believed that complicated motions of molten iron in the core generate electric currents by interacting with a complicated magnetic field. The resulting dipole-like portion of the field is detected on the Earth's surface. However, details of the motions of molten iron and the magnetic fields in the core are not known. In

Figure 1b. The aurora seen from the ground (courtesy of the Kanazawa Astronomical Society).

spite of such difficulty in understanding the generation mechanism of the Earth's magnetic field, it is interesting that it can be described to a high degree of accuracy in terms of a bar magnet located near the center of the Earth.

It is this magnetic field and the Earth's atmosphere that protect life from the hot solar wind. In fact, Earth's magnetic field acts as a barrier against the charged particles of the solar wind. As a result, the solar wind is deflected around the Earth, forming a comet-shaped cavity (Figure 3). This cavity is called the magnetosphere [*Cowley*, this vol.], and its dayside boundary is located at a distance of about 10 Earth radii.

Like Earth, the Sun is also a magnetized celestial body [*Hathaway*, this vol.]. As the solar wind blows out from the Sun, it carries away a part of the magnetic field lines by stretching them like a bundle of rubber bands. As solar wind magnetic field lines reach the boundary of the magnetosphere, they interact with those of Earth's magnetic field, interconnecting themselves [*Cowley*, this vol.]. The scientific term for this process is "reconnection," because the process also involves "disconnection" to begin with. As a result, some of the solar magnetic field lines and Earth's magnetic field lines interconnect across the boundary of the magnetosphere (Figure 3). Those

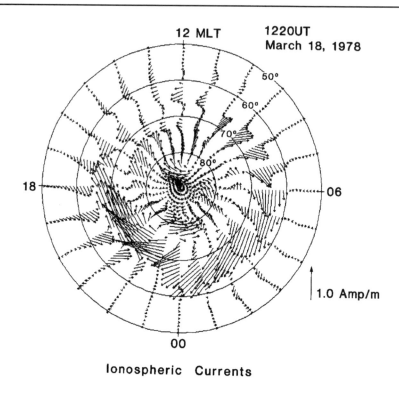

Ionospheric Currents

Figure 2. The distribution of electric currents in the polar ionosphere (seen from above the northern geomagnetic pole). Note concentrated currents along the auroral oval, particularly in the night sector.

field lines are called "open" field lines. The Earth's magnetic field lines involved in this process originate from the area surrounded by the auroral oval, which is approximately centered at the geomagnetic pole (Figure 4). The radius of the oval area is about 2500 km.

Since the solar wind blows along the boundary of the magnetosphere and thus across the reconnected field lines, this process constitutes a generator, in which an electrical conductor (the free electrons and protons in the solar wind) moves in a magnetic field. The solar wind-magnetosphere generator can generate more than 1,000,000 MW, the induced voltage being about 20–150 kV. The magnetohydrodynamic (MHD) generator, a laboratory and industrial device used to generate electricity by forcing an ionized gas (plasma) through a magnetic field, works on the same principle.

In a rarefied plasma permeated by a magnetic field, charged particles can move freely only along magnetic field lines. For this reason, the mag-

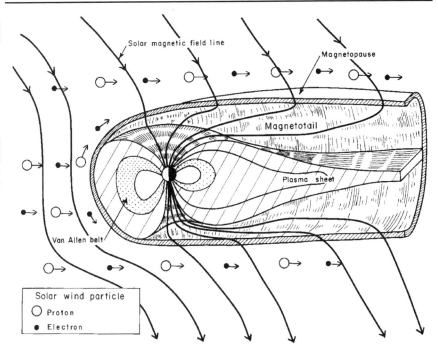

Figure 3. A schematic diagram showing the noon-midnight cross-section of the magnetosphere, and some important internal structures. Some of the solar wind magnetic field lines, the Earth's magnetic field lines, and reconnected field lines are shown. Note that the solar wind particles flow across the reconnected field lines. The insert shows the primary discharge circuit powered by the solar wind-magnetosphere generator.

netic field lines act like a conducting wire to carry electric currents. As mentioned earlier, the reconnected field lines originate from a circular area approximately centered at the geomagnetic pole (Figure 4). Though details will be omitted, the "terminals" of the solar wind-magnetosphere generator are connected to the boundary of this oval area by magnetic field lines. Therefore, the discharge process powered by the generator takes place between the morning side of the magnetosphere boundary (the positive "terminal"), goes through the morning half of the boundary of the oval area in the polar region of the Earth, then mostly along the auroral oval, and finally out from the evening half of the oval area's boundary to the evening side of the magnetospheric boundary (the negative "terminal"); see the insert in Figure 3. This system is the primary discharge circuit and is called the region 1 current system. The region 1 current system induces a secondary current system, called the region 2 current system, which is the part

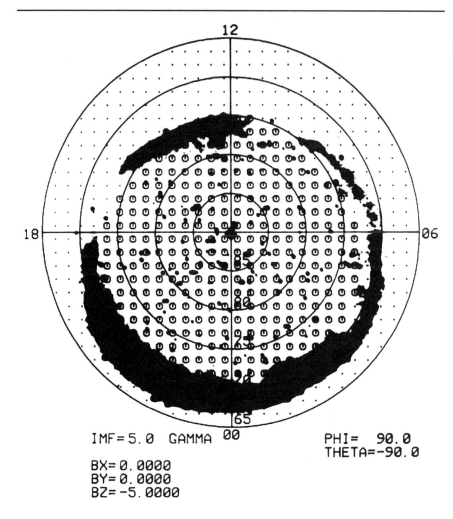

Figure 4. An image of the auroral oval. The 'feet' of the 'open' geomagnetic field lines are shown by a dot with a circle. The feet of the other field lines ('closed' field lines) are shown by a dot. Note that the area of the feet of the open field lines are surrounded by the auroral oval.

of the region 2 current that is discharged back to the magnetosphere from the "equatorward" boundary of the auroral oval in the morning side and from the magnetosphere to the equatorward boundary of the auroral oval in the evening side. Electric currents flowing along magnetic field lines are called field-aligned currents. The currents are mostly carried by electrons, so there is a downward flow of electrons carrying an upward current from the

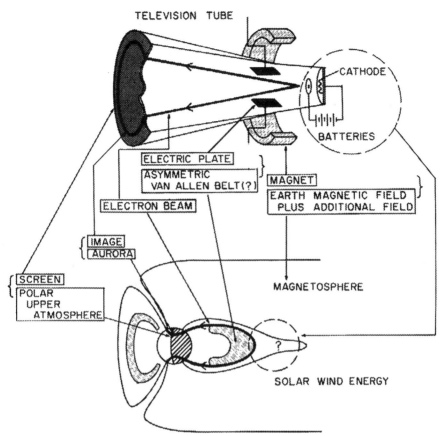

Figure 5. A schematic diagram indicating a close analogy between a cathode ray tube and the magnetosphere. The polar upper atmosphere corresponds to the screen of the tube, while the aurora corresponds to an image on the screen.

magnetosphere to the ionosphere along the poleward boundary of the auroral oval in the evening side and along the equatorward boundary of the oval in the morning sector.

These downward flowing electrons deposit their energy in the upper atmosphere by exciting and/or ionizing atoms and molecules. Some of this energy is released in the form of visible light, which we recognize as the aurora. Although not well understood at present, the downward streaming electrons form one or more thin sheet beams, electrical currents flowing in thin layers, as opposed to thin filaments. As a result, the aurora has the form of a curtain (Figure 1b). The most common color of the aurora is greenish-white, due mainly to the green from the excited oxygen atoms at a wave-

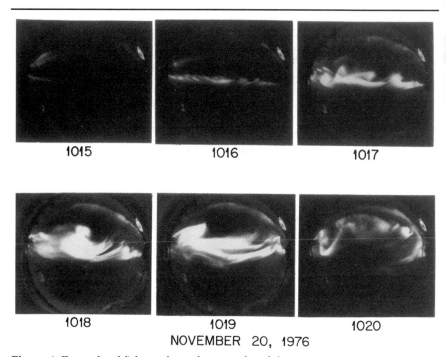

1015   1016   1017

1018   1019   1020
NOVEMBER 20, 1976

Figure 6. Example of fish-eye lens photographs of the aurora at substorm. The top of each circular photograph is the northern horizon. The photographs were taken at the Poker Flat Research Range, Geophysical Institute, University of Alaska, Fairbanks.

length of 5577 Å or 5577 nm. At auroral altitudes, which are greater than 100 km, a significant part of the oxygen molecules are dissociated into atoms. Active "auroral curtains" often show a splendid pinkish or magenta color at their lower borders. This particular color arises from molecular nitrogen band emissions.

There are at least two major mysteries involved in the process of sheet beam and auroral curtain formation. First, the north-south thickness of an auroral curtain is only about 1 km or less, although an auroral curtain stretches a distance of several thousand kilometers along the auroral oval. Often, as many as five or more curtains appear along the oval. There is no satisfactory explanation for the formation of these thin layers. Secondly, the bottom height is about 100 km in altitude, ten times as high as the cruising altitude of a commercial jet aircraft. The upper height of the curtain is diffuse, and extends at least up to 400–500 km in altitude. In order for the electrons to reach an altitude of 100 km, they must have energy of at least a few kiloelectron volts. Since the energy of electrons in the magnetosphere is only a few hundred electron volts, they must be accelerated to such energies before they arrive at the top of the atmosphere. Although there have been many ideas put forward to explain these features, once again, there is no widely accepted theory at this time.

The field-aligned currents close by being connected to horizontal currents in the ionosphere [*Richmond*, this vol.]. The main horizontal currents are concentrated at two places along the oval; a westward current in the morning side of the oval and an eastward one in the evening side (Figure 2). These two currents are commonly of the order of 1,000,000 amp and are called the auroral electrojets. A significant amount of the power created by the solar wind-magnetosphere generator is dissipated in terms of Joule heat energy in the ionosphere, since the ionosphere is a resistive medium for the currents. A large-scale atmospheric circulation of the upper atmosphere is generated by the heating process. Theoretical estimates of atmospheric heating by the aurora have been made and incorporated in thermospheric circulation models, but experimental measurements of these processes are difficult and sparse.

Auroral curtains have very complicated motions in which curls of various scales occur that were once called "the horseshoe type" or "the drapery type." The smallest scale curls are like pleats. Some of them are produced because the electron sheet beam is negatively charged, so a pair of oppositely-directed electric field develop across the curtain. Such a pair of electric fields causes a counterflow of the ionospheric plasma along the curtain, resulting in a series of eddies. When they are viewed perpendicular to the curtain, they appear as parallel "rays." These rays appear to move rapidly, 20 km per second, to the east and west along the arc.

Auroral curtains move rapidly northward or southward. These motions are not caused by moving light-emitting atoms and molecules. The northward motion is caused by the electron sheet beam shifting northward so that its impact point in the ionosphere also shifts northward. Indeed, there is a good analogy between an image in a television and the aurora (Figure 5). The aurora corresponds to an image on the screen of the tube. Motions of the image are produced by motions of the impact point of the electron beam on the screen, which are, in turn, produced by changes of the electric and magnetic fields in the modulation devices in the tube. These are produced by input changes. For this reason, many auroral physicists are trying to infer changes of the electric and magnetic fields (or electric/magnetic storms) in the magnetosphere on the basis of observed auroral motions. These changes result from changes of the input, namely of the solar wind and the solar magnetic field in the solar wind.

In spite of the great complexity involved in auroral motions, there is a systematic aspect to them. Auroras may be seen undergoing a global-scale activity called the auroral substorm. At the onset of an auroral substorm, an auroral curtain in the midnight sector suddenly increases it brightness by an order of magnitude or more, and starts to move poleward with a speed of a few hundred meters per second (Figure 6). Thus, the thickness of the midnight part of the oval increases rapidly producing a large bulge, which in turn produces a large—scale wavy structure in the late evening sector. This particular feature propagates along the auroral oval toward the dusk twilight sector and is descriptively called the westward traveling surge. The wavy structure is a very spectacular display. At the same time, auroras in the morning sector appear to disintegrate into many rays. This activity lasts for about 2–3 hours, and the aurora all along the oval becomes quiet. An auroral substorm repeats several times in a moderately disturbed day.

The auroral substorm is caused by a tenfold increase of the power of the solar wind-magnetosphere generator, from about 0.1 million MW during a quiet period to a speed of about 1 million MW for a few hours. The direction of the solar wind magnetic field plays a crucial role in this increase. If the solar wind magnetic field has a southward-directed component, a larger number of both the solar wind magnetic field lines and Earth's magnetic field lines interconnect, increasing the intensity of the magnetic field across the boundary of the magnetosphere. The increased interconnection is equivalent to an increase of the magnetic field in a generator.

After a major flare on the Sun [*Rust*, this vol.], the associated solar wind is intensified and also carries a stronger solar wind magnetic field. A result can be a hundredfold increase of the generating power if the strong wind is directed toward Earth—or if the flare occurs near the center of the visible solar disk—and if the magnetic field is directed southward, opposite to the Earth's magnetic field at the equatorward boundary of the magnetosphere. Such an increase of the power will cause stronger discharge currents and brighter auroras. The discharge current produces intense, rapidly varying magnetic fields that we identify as the geomagnetic storm field [*Tsurutani and Gonzalez*, this vol.]. This is why intense auroral activity is accompanied by an intense geomagnetic storm that causes complicated problems in long distance power transmission lines, oil pipelines, communication lines, and some defense radar systems, etc. [*Joselyn*, this vol.].

It is possible to test what is described above as the answer to the question "What causes the aurora?" Three elements, the solar wind, the magnetic field, and the atmosphere of a planet or its satellites, are needed to create the aurora. All of the planets are exposed to the solar wind. Mercury has an Earth-like magnetic field, but no atmosphere. Thus, there is no aurora. Both Venus and Mars do not have a detectable magnetic field, and there is no aurora on either planet. The Moon has no magnetic field, no atmosphere to speak of, and thus no aurora. Jupiter has a very strong magnetic field and an atmosphere consisting mainly of hydrogen. Since Jupiter is much larger than

the Earth, the Jovian auroral oval is much larger than Earth. The Jovian aurora is visually pinkish or magenta, resulting from hydrogen atom emission.
Saturn also has an intense magnetic field and an atmosphere. The aurora on
Saturn was recently imaged. Both Uranus and Neptune have a magnetic
field an atmosphere. The aurora on Uranus was imaged by the Voyager
spacecraft.

A study of the aurora involves the generation of electric power and the
subsequent discharge process in the natural plasma environment of the
Earth's magnetosphere and ionosphere. A similar environment may exist in
the solar atmosphere, stars, pulsars, etc. For this reason, some auroral physicists believe that the solar flares are perhaps a sort of "solar aurora"; on the
other hand, some solar physicists believe that the auroral substorm is an
"Earth flare." A good understanding of the auroral phenomena is the foundation in understanding many astrophysical problems, including solar
flares.

Syun-Ichi Akasofu
Geophysical Institute, University of Alaska, Fairbanks, AK 99775–0800.

The Earth's Magnetosphere

S. W. H. Cowley

The Earth's magnetosphere is composed of two essential ingredients. The first is the Earth's magnetic field, generated by currents flowing in the Earth's core. Outside the Earth this field has the same form as that of a bar magnet, aligned approximately with the Earth's spin axis. The second ingredient is the solar wind [*Goldstein*, this vol.], a fully ionized hydrogen/helium plasma that streams continuously outward from the Sun into the solar system at speeds of ~300–800 km s^{-1}. This plasma wind is pervaded by a large-scale interplanetary magnetic field (IMF), which plays a crucial role in the Earth's interaction with the solar wind. There is also a third ingredient that plays an important role: The Earth's ionosphere [*Richmond*, this vol.; *Sentman and Wescott*, this vol.]. The upper atmosphere is partially ionized by solar far-ultraviolet and X-rays above altitudes of ~100 km. The ionosphere forms a second source of plasma for the magnetosphere, mainly of protons and singly charged helium and oxygen.

The Chapman-Ferraro Magnetosphere

The basic nature of the interaction between the solar wind and the Earth's magnetic field was first deduced by Chapman and Ferraro in the early 1930's. It is based on two theoretical principles. The first concerns the way in which plasmas and magnetic fields interact; they behave, approximately, as if they are "frozen" together. This follows as a consequence of Faraday's law, from the fact that in an electrically conducted plasma the electric field in the rest frame must be close to zero, otherwise very large electric currents would be driven. As a result of this freezing together, magnetic fields are transported by flowing plasmas; the field lines are bent and twisted as the flow bends and twists. An important example is the IMF mentioned above, which represents the solar magnetic field transported outwards into the solar system by the solar wind. It is wound into a large spiral structure by the Sun's rotation [*Goldstein*, this vol.], and near the Earth has a strength of ~5 nT. The second principle concerns the magnetic field's effect on the plasma, which arises from the Lorentz force $q\mathbf{V}\times\mathbf{B}$ experienced by a charge q moving with velocity $\mathbf{V}$ in a magnetic field $\mathbf{B}$. Summed over all the charges in a given region, the net force usually opposes the bending and twisting of the field, or its compression, in the frozen-in flow. There are two components of this force. First, the field exerts an effective pressure on the plasma proportional to the square of the magnetic field strength. This force resists compressions or rarefactions of the magnetic field. Second, bent field lines exert a tension force on the plasma, like that of stretched rubber bands. This force resists the bending and twisting of field lines.

Applying these ideas to the interaction between the solar wind and the Earth, we conclude with Chapman and Ferraro that since the solar wind plasma is frozen to the IMF, and the Earth's plasma to the Earth's field, the plasmas will not mix. Instead, the solar wind will confine the Earth's field to a cavity surrounding the planet, forming a magnetosphere (Figure 1a). The size of the cavity is determined by pressure balance at the boundary between the pressure of the solar wind on one side, and the magnetic pressure of the planetary field on the other. Given a planetary "bar magnet" (dipole) field that produces a field strength of ~30,000 nT at the Earth's surface at the equator, estimates place the boundary, the magnetopause, at a geocentric distance of ~10 R_E on the upstream side, and this where it is generally observed. R_E is the Earth's radius, equal to ~6400 km. On the downstream side the cavity extends into a long magnetic tail whose form is determined by additional physics to be outlined below. Across the magnetopause the magnetic field usually undergoes a sharp change. Ampére's law then tells us that a sheet of electrical current must flow in the plasma in this interface. A bow shock also stands in the solar wind upstream of the cavity (Figure 1a), which forms because the speed of the solar wind relative to the Earth is much faster than that of wave propagation within it. Across the shock the flow is slowed, compressed, and heated, forming a layer of turbulent plasma outside the magnetopause called the magnetosheath. Inside the cavity, in this simple picture, the terrestrial plasma roughly corotates with the Earth (Figure 1a). This occurs because the Earth's field lines are frozen into the ionospheric plasma, where approximate corotation with the Earth is enforced, in the absence of other driving processes, by collisions between ions and atmospheric neutrals in the upper E-region (at heights of ~120–140 km).

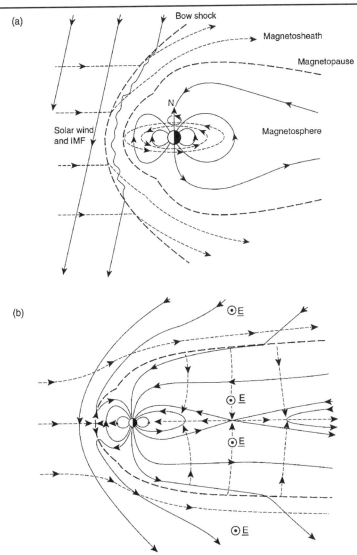

Figure 1. Sketches of the structure of the Earth's magnetosphere in the noon-midnight meridian plane, showing (a) the Chapman-Ferraro closed magnetosphere based on strict application of the frozen-in approximation, and (b) the Dungey open magnetosphere, in which there is an essential breakdown of frozen-in flow at the dayside magnetopause and in the tail leading to the occurrence of reconnection. The arrowed solid lines indicate magnetic field lines, the arrowed dashed lines plasma streamlines, and the heavy long-dashed lines the principal boundaries (the bow shock and magnetopause). The circled dots marked *E* in (b) indicate the electric field associated with the flow, which varies in strength in different locations.

Dungey's Open Magnetosphere

In Figure 1a the interplanetary magnetic field lines are compressed against the magnetopause and draped over it by the flow, but ultimately slip around the "sides" of the magnetosphere, frozen into the magnetosheath plasma. However, the "frozen-in" picture is only an approximation, and under some circumstances it will break down. One of those circumstances occurs when high current densities are present in the plasma, as occurs in Figure 1a at the magnetopause. Dungey, in the early 1960's, was the first to recognize the importance of this breakdown and to study its consequences. When the frozen-in condition is relaxed, the field will diffuse relative to the plasma in the magnetopause, allowing the interplanetary and terrestrial field lines to connect through the boundary (Figure 1b). Dungey called this process magnetic reconnection. The distended loops of "open" magnetic flux formed by reconnection exert a magnetic tension force that accelerates the plasma in the boundary north and south away from the site where reconnection takes place, thus causing the open tubes to contract over the magnetopause toward the poles. This flow was first observed by the ISEE-1 and -2 satellites in 1978. The open tubes are then carried downstream by the magnetosheath flow, and stretched into a long cylindrical tail. Eventually the open tubes close again by reconnection in the center of the tail. This process forms distended closed flux tubes on one side of the reconnection site, which contract back toward the Earth and eventually flow to the dayside where the process can repeat. On the other side, "disconnected" field lines accelerate the tail plasma back into the solar wind. The key feature of the "open" magnetosphere is therefore the cyclical flow excited in the interior by these reconnection processes. Figure 1b shows that this flow is associated with a large-scale electric field in the plasma directed from dawn-to-dusk across the system given by $E = -V \times B$ (equivalent to the statement that E in the plasma rest frame is zero). From Faraday's law the voltage across the system associated with this electric field is equal to the total magnetic flux throughput, and represents a measure of the overall strength of the flow. This voltage cannot be measured directly in the magnetosphere because of its vast spatial scale, but it can be determined at ionospheric heights using flow data from ground-based radars or low-altitude spacecraft. The image projected onto the ionosphere of the magnetospheric flow is shown in Figure 2. It consists of twin vortices with antisunward flow of open field lines over the polar cap and a return sunward flow of closed field lines at lower latitudes. The voltage between the "foci" of the vortices is of order 100 kV, and is associated with ionospheric flows of several hundred m s^{-1}. The overall flow cycle time may also be estimated from ionospheric flow measurements, and turns out to be ~12 h, of which field lines remain open mapping into the tail lobe for ~4 h, and

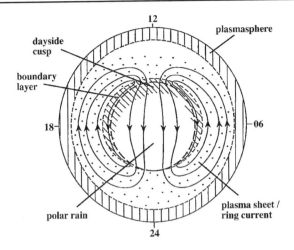

Figure 2. View looking down on the northern high-latitude ionosphere showing the plasma convection streamlines (arrowed solid lines) and the main zones of magnetospheric plasma precipitation. The interior heavy dashed circle shows the boundary between open field lines at high latitudes and closed field lines lower latitudes.

then take ~8 h to convect back from the tail to the dayside. The ~4 h interval for which a field line remains open, combined with the speed of the solar wind, allows us to estimate the length of the tail lobes as ~1000 R_E.

The most important feature of the magnetosphere-ionosphere flow, however, is that its strength is modulated by variations in the IMF. The dayside reconnection rate, and hence the flux throughput in the magnetosphere, is strong when the IMF points south, opposite to the equatorial field of the Earth (Figure 1b). When the IMF points north, however, equatorial reconnection cannot occur, and the flow dies away. This dependence of the flow on the direction of the IMF distinguishes Dungey's open model from other possibilities (for example, flow excited in a closed Chapman-Ferraro system by "viscous" stresses at the magnetopause), and has been demonstrated by many studies over the past 30 years. It is important to realize, however, that the magnetic flux throughput in the system, even at its strongest, amounts to no more than about 20% of the interplanetary flux brought up to the dayside magnetosphere by the solar wind. Most of the interplanetary flux is indeed deflected around the magnetosphere as deduced by Chapman and Ferraro. However, it is the breakdown of this picture at the 10–20% level that is critical to the Earth's magnetospheric dynamics. The contribution to the flux transport by other non-reconnection mechanisms appears to be smaller than this by roughly a factor of ten.

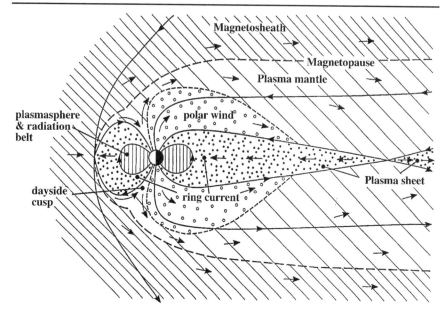

Figure 3. Sketch showing the principal plasma populations in the Earth's magnetosphere in a view in the noon-midnight meridian plane. The solid arrowed lines indicate magnetic field lines, the heavy long-dashed line the magnetopause, and the arrows the direction of plasma flow. Diagonal hatching indicates solar wind/magnetosheath plasma and populations derived directly from it, open circles indicate outflowing ionospheric plasma, solid dots the hot plasma accelerated in the tail, and the vertical hatching the corotating plasmasphere. These indicators are the same as those employed in Figure 2.

Magnetospheric Plasma Populations

A basic sketch of magnetospheric plasma populations is shown in Figure 3. Beginning on the dayside, the production of open flux tubes at the magnetopause provides a direct magnetic path along which both warm magnetosheath plasma (100 eV–1 keV) can escape. The magnetosheath plasma flows down the newly opened field lines toward the Earth, forming the "dayside cusp" population, where it is repelled by the increasing field strength via the "magnetic mirror" effect. Most plasma is reflected back along the open field lines away from the Earth because of this effect, but due to the flow over the pole, the reflected plasma flows up into the outer part of the tail, forming the plasma mantle population, where it is augmented by continual inflow of magnetosheath plasma across the tail lobe magnetopause. As the mantle plasma flows down-tail, it is convected toward the tail center plane at down-tail distances beyond 100–200 R_E. The inner part of the open tail is populated by low-energy (10–100 eV) streams of protons and singly charged oxygen ions from the ionosphere (marked "polar wind" in Figure 3). Due to the low speed of these ions (a few tens of km s^{-1}), they are convected into the center plane relatively close to the Earth, typically within ~100 R$_E$.

When it reaches the center plane, the tail lobe plasma is accelerated to speeds of 500–1000 km s^{-1} (proton energies of ~1–5 keV) by the tension of the field lines downstream from the reconnection site, forming the plasma sheet population. The tail reconnection site usually lies ~100 R$_E$ from Earth, but it comes closer during periods of magnetic disturbance. The plasma that flows into the center plane tailward of the reconnection site is accelerated away from the Earth and eventually rejoins the solar wind. The plasma accelerated earthward of the reconnection site is transported on closed flux tubes toward the Earth. Individual particles bounce from hemisphere to hemisphere because of the magnetic mirror effect. Additional "magnetic drifts" cause the particles to drift across the tail at a rate proportional to their energy, ions in one direction and electrons in the other; this population carries the current in the tail center required by Ampére's law. Overall, the plasma sheet population becomes denser and hotter as it convects toward the Earth, due to the compression of the plasma on the contracting closed flux tubes. As this hot plasma enters the inner quasidipolar magnetosphere and convects toward the dayside, it becomes known as the ring-current plasma [*Van Allen*, this vol.], because the drifting particles carry an electric current westward around the Earth.

The hot population does not penetrate all the way to the Earth in the equatorial plane, however, because of the tendency of the plasma, in the absence of other driving mechanisms, to corotate with the Earth. When the flows caused by the Dungey cycle are added to the flows from corotation, the former dominate in the outer part of the magnetosphere while the latter are confined to a central core of dipolar flux tubes. This corotating core extends, on average, to distances of 4–5 R$_E$ in the equatorial plane. In the steady state the inner corotating region is filled to high densities (10^2–10^3 cm^{-3}) with cold (~1–10 eV) hydrogen/helium plasma from the topside ionosphere, forming the plasmasphere. Outside this region the cold plasma density is much lower (0.1–1 cm^{-3}) because of heating and loss of the ionospheric plasma during each convection cycle, and the plasma is instead characterized by the presence of the hot (10–100 keV) and tenuous (~1 cm^{-3}) ring current plasma originating in the tail. The size of the plasmasphere is not constant, however. The corotation region extends further from Earth than average during intervals of weak convection, and the ionosphere fills the outer flux tubes toward equilibrium values while this condition persists. The corotating region shrinks when the convective flow increases again, and the cold plasma that accumulated in the outer region is stripped away and flows to the dayside magnetopause. It is replaced by hot plasma flowing in from the tail.

Magnetospheric Substorms

The variability of the magnetospheric flow on time scales of minutes and hours which is associated with changes in the direction of the IMF has been mentioned above as a key feature. However, observations show that when dayside reconnection is enhanced by a southward turn of the IMF, the magnetosphere generally does not evolve smoothly toward a new steady state of enhanced convection. Instead, the system, particularly the tail, undergoes a characteristic evolution on a 1–2 h time scale called a magnetospheric substorm [*Akasofu*, this vol.].

Suppose the magnetosphere is initially in a state of low flow during an interval of northward IMF, and that the IMF turns south. Reconnection starts at the magnetopause, stripping flux off the dayside and adding it to the tail, so that the dayside magnetopause moves in (by up to ~1 R_E), while in the tail both the radius and the field strength increase. This change is accompanied by an excitation of large-scale flow on 10–20 min time scales as the system adjusts. As the tail develops, the current in the near-Earth (~10 R_E) portion of the plasma sheet becomes concentrated in a layer that is only 500–1000 km thick (Figure 4a), compared with quiet-time thicknesses of ~30,000 km. This thin layer develops in intensity but otherwise remains stable during this initial "growth phase," which lasts tens of minutes. Why this happens is not understood, and it is the subject of much research. Then, on time scales of ~1 min, the layer disrupts, again for reasons yet to be determined. The current suddenly decreases and the distended tail-like field lines collapse inward toward the Earth near the equator, and outward at higher latitudes, to a more dipolar form (Figure 4b). As they do so, the plasma they contain is strongly heated and compressed. This produces a sudden and intense flux of keV electrons at the top of the atmosphere at the feet of these field lines, producing a brilliant auroral display as the atmospheric atoms are excited and radiate. The collapse usually starts in a restricted longitude sector near midnight in the near-Earth end of the tail (typically at distances of 8–12 R_E), and then propagates both down and across the tail. In the atmosphere the area of bright auroras expands poleward and to the east and west, such that this interval is known as the expansion phase of the substorm. This field collapse often (but perhaps not invariably) induces the onset of reconnection in the plasma sheet at distances of 20–40 R_E (Figure 4c) as it propagates down-tail. When this happens, much of the plasma sheet can be "pinched off" to form a closed-loop

(a)

(b)

(c)

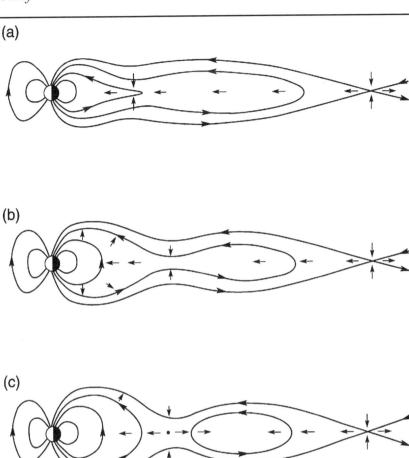

Figure 4. The tail plasma sheet is shown developing during the growth and expansion phases of a magnetospheric substorm. Sketch (a) shows the development of a thin current layer at the center of the near-Earth tail during the growth phase, (b) shows the onset of current disruption and field dipolarization, and (c) shows the induction of tail reconnection and plasmoid formation at larger distances from Earth.

plasmoid that propagates down-tail and into the solar wind at speeds of 400–800 km s^{-1}. Continued reconnection in the near-Earth tail then closes open flux in the tail lobe. After a few tens of minutes, however, the reconnection rate slackens and the reconnection region moves back down the tail, signaling the end of expansion. The subsequent recovery phase typically lasts for many tens of minutes. The system responds and adjusts to the input of magnetic flux and hot plasma on the nightside during the expansion phase during this time.

The above sequence of events typically follows a southward turning of the IMF which endures for at least a few tens of minutes. If the southward field persists significantly longer, however, the system often evolves through a series of substorm cycles each lasting ~1 h. The hot ring current plasma is then driven deep into the inner magnetosphere by the enhanced convection, constituting the main characteristic of a magnetic storm [*Tsuratani and Gonzalez,* this vol.]. The westward current carried by this plasma produces a world-wide decrease in the field strength on the ground at low latitudes whose amplitude may reach several hundred nT.

Much has been learned about the Earth's magnetosphere during the past 30 years, but a number of uncertainties remain. While we do not yet have all of the answers, it is clear that the questions we are asking have considerably improved.

Further Reading: An extended introductory treatment of these topics may be found in *Introduction to Space Physics,* edited by M. G. Kivelson and C. T. Russell (Cambridge University Press, 1995). This book also contains numerous references to the relevant research literature. Additional introductory material may be found in *Sun and Earth,* by H. Friedman (Scientific American Books, Inc., 1986), and in *The Solar-Terrestrial Environment,* by J. K. Hargreaves (Cambridge University Press, 1992).

S.W.H. Cowley
Dept. of Physics, University of Leicester, University Road, Leicester LEI 7RH
United Kingdom

Radiation Belts

James A. Van Allen

In this article, I describe some of the reasons for the existence of radiation belts around the Earth and other planets and some of their basic properties. A radiation belt is an important component of a larger and more complex physical system called a magnetosphere [*Cowley*, this vol.]. Elsewhere the similarities and differences among the diverse and dynamic planetary magnetospheres that have been investigated thus far are discussed in a more general context [e.g., *Tsurutani and Gonzalez*, this vol.; *Akasofu*, this vol.].

A magnetosphere is that region surrounding a planet within which the planet's intrinsic magnetic field has an important role in physical phenomena involving electrically charged particles. But even unmagnetized planets, satellites of planets, comets, and (presumably) asteroids exhibit similar rudimentary plasma physical effects. The Earth's magnetosphere extends about 10 planetary radii toward the Sun and hundreds of times that far in the direction away from the Sun. Its outer boundaries and much of its physical dynamics are attributed to the solar wind—the tenuous, ionized, magnetized gas (plasma) that flows outward from the solar corona through interplanetary space. The solar wind [*Goldstein, this vol.*] does not readily penetrate the geomagnetic field but compresses and confines the field around the Earth. The sunward boundary is located where the external pressure of the flowing solar wind equals the internal pressure of the geomagnetic field. A complex process of interconnection of the solar wind's magnetic field and the geomagnetic field stretches out the magnetic field in the direction away from the Sun, creating the long magnetotail.

A radiation belt is an interior feature of a magnetosphere and comprises a population of energetic, electrically charged particles (electrons, protons, and heavier atomic ions) durably trapped in the magnetic field of the planet. In this context the term energetic conventionally means kinetic energies $E \geq 30$ kiloelectron Volts (keV). A radiation belt is toroidally shaped, encircles the planet, and its axis of rotational symmetry is coincident with the magnetic dipolar axis of the planet. To a first approximation, each particle therein moves with constant energy and independently of all other particles along a helical path encircling a magnetic line of force. This motion is subject only to the Lorentz force of a static magnetic field on a moving electrically charged particle, namely $q(\mathbf{v} \times \mathbf{B})/c$, where q is the particle's electrical charge, $\mathbf{v}$ its velocity, c the speed of light, and $\mathbf{B}$ the local magnetic field. The angle between $\mathbf{v}$ and $\mathbf{B}$ (the pitch angle of the helix) tends toward either $0°$ or $180°$ at the magnetic equator during each latitudinal excursion and becomes $90°$ at mirror or reflection points in the northern and southern hemispheres as the particle penetrates into the stronger magnetic field near the planet. The helix drifts slowly in longitude, westward for $q > 0$ (protons and other ions) and eastward for $q < 0$ (electrons), so as to generate the overall toroidal shape of the trapping region.

In this simplified, idealized case of motion in a vacuum in a dipolar magnetic field, each particle has an infinite residence time. All of this was shown theoretically in 1907 by Störmer, but he did not suggest any geophysical significance of his findings. Departures from the Störmerian model in real magnetospheres are caused by the presence of thermal and quasi-thermal ionized gas (plasma) that causes a large variety of cooperative physical phenomena, essential to understanding magnetospheric dynamics. In analyzing such processes the Störmerian approach is usually supplanted by characterizing a particle's motion in the Earth's magnetic field with three adiabatic invariants, corresponding to the three cyclic components of motion having widely different periods, namely, gyration around (~milliseconds), latitudinal oscillation (~seconds), and longitudinal drift (~hours).

The magnetospheric properties of a planet are an essential part of its gross phenomenological character. They define the external environment and reflect the internal properties of the planet. The energetic particle population may place important constraints on the practicality of in situ measurements and on the survival of electronic and optical equipment, human flight crews, animals, and other life forms flown therein [*Joslyn*, this vol.]. The particle population of the Earth's radiation belts makes it dangerous for humans without massive shielding to do more than quickly pass through them. In an extreme example, the Jovian radiation belts are sufficiently intense to damage some solid state electronics, even during brief flythroughs.

In contrast to popular misunderstandings, natural radiation belts are not composed of radioactive nuclei, nor does the population of energetic particles shield the planet from external radiations. The "radiation" is comprised of the trapped energetic particles themselves. There is a large amount of shielding from external radiations at the surface of the Earth, but this shielding is provided by the atmosphere and, to a lesser degree, by the geomagnetic field.

The diverse particle phenomena in the Earth's magnetic field have been studied intensively, both observationally and theoretically, since my discovery of their existence in 1958. In addition, a series of artificial radiation belts were produced by the United States and the Soviet Union in 1958 and 1962. The energetic particles (principally electrons) in these artificial belts were the decay products of radioactive fission nuclei injected into the magnetic field by nuclear bomb bursts at high altitudes. Other active experiments are of increasing importance.

The entire body of knowledge of both natural and artificial radiation belts and the associated plasma physical phenomena at the Earth define the prototypical planetary magnetosphere.

The physical mechanisms for the creation of magnetospheric phenomena are of an electromagnetic nature. Within the solar system, the minimum condition for the existence of a planetary radiation belt is that the planet's dipole magnetic moment be sufficiently great that the flow of the solar wind is arrested before it reaches the top of the appreciable atmosphere or surface of the planet. Durable trapping of charged particles is possible only if this condition is met. Otherwise, particles are lost quickly by collisions with atmospheric gas or the solid body of the planet. But even when the foregoing condition is not met, important plasma physical phenomena still occur, and have been observed near the Moon, Mercury, Venus, Mars, and the comets Giacobini-Zinner and Halley.

Beyond the elementary consideration mentioned above, there are very complex physical processes of thermalization and capture of the solar wind (plasma), convective transport of it by the combination of the planetary magnetic field and induced electric fields, and the acceleration and diffusion of charged particles by fluctuating electric and magnetic fields—all of which contribute to the development of the total magnetospheric system. A reasonable level of understanding of these processes at the Earth has been achieved and certain scaling principles can be used as guidelines for interpreting the magnetospheres of other planetary bodies. Of the eight planets investigated, only Jupiter, Saturn, Uranus, Neptune, and Earth have well-developed radiation belts.

The radiation belts and other features of the Earth's magnetosphere are shown to approximate scale in the noon-midnight meridian plane cross section of Figure 1. The inner and outer radiation belts are two distinct features, defined by the intensity of particles capable of penetrating a specific shield (~1 g cm^{-2} of aluminum). In a generalized sense, there are as may different radiation belts as there are different species of particles and energy ranges that one wishes to distinguish. The principal sources of particles for the outer belt are the solar wind and the ionosphere; for the inner belt,

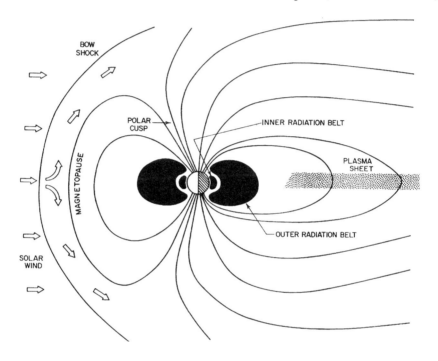

Figure 1. Principal features of Earth's magnetosphere to approximate scale in the noon-midnight meridian plane cross section.

important additional sources is electrons and protons from the in-flight radioactive decay of neutrons from nuclear reactions produced by galactic cosmic rays [*Jokipii*, this vol.) and solar energetic particles [*Lin*, this vol.). The eventual fate of trapped magnetospheric particles is to become part of the atmosphere, to collide with satellites or particulate matter in planetary rings, or to escape into space. The first two sources of particles are responsible for most of the gross geophysical manifestations of the magnetosphere: aurorae [*Akasofu*, this vol.], geomagnetic storms [*Tsurutani and Gonzalez*, this vol.], and heating of the upper atmosphere. The third is responsible for the relatively stable population of very energetic protons and some of the energetic electrons in the inner radiation belt. It is noted that this third source would produce a radiation belt around a magnetized planet even if the solar wind did not exist.

The residence times of individual particles in the radiation belts of Earth, controlled by ionization losses in the atmosphere near the Earth (altitudes of < 400 km), increase rapidly to the order of years at a radial distance of about 8,000 km (1.25 Earth radii), then decline in a complex and time-variable way to values of the order of weeks, days, and minutes in the outer fringes. There are quite low intensities of radiation-belt particles within a spherical shell of about 400-km thickness around Earth. This is the region of space flight that is relatively safe from the radiation point of view. The inner radiation belt extends from this lower boundary to an equatorial radial distance of about 12,000 km and the outer radiation belt from this point outward to about 60,000 km. There is considerable overlap of the two principal belts and a complex and time-variable structure in the outer one. Some sample omnidirectional intensities are $J = 2 \times 10^4$ (cm^2 s)$^{-1}$ of protons $E_p > 30$ MeV in the most intense region of the inner belt; and $J = 3 \times 10^8$ (cm^2 s)$^{-1}$ of protons $E_p > 0.1$ MeV, $J = 2 \times 10^8$ (cm^2 s)$^{-1}$ of electrons $E_e > 0.04$ MeV, and $J = 1 \times 10^4$ (cm^2 s)$^{-1}$ of electrons $E_e > 1.6$ MeV in the most intense region of the outer belt.

Great advances in knowledge of planetary radiation belts and magnetospheres have been achieved by appropriately instrumented spacecraft in their close encounters with Jupiter in 1973 (Pioneer 10), 1974 (Pioneer 11), and 1979 (Voyagers 1 and 2) with Saturn in 1979 (Pioneer 11), 1980 (Voyager 1), and 1981 (Voyager 2); with Uranus in 1986 (Voyager 2), and with Neptune in 1989 (Voyager 2). These four planets all have well-developed magnetospheres and radiation belts, with certain basic features in common but with distinctive features in each case. For example, the highest radiation belt intensities seen at Earth are much less than those found at Jupiter, Saturn, Uranus, or Neptune. At Earth, the ionic component of the radiation belts is comprised mainly of protons with small ad-mixtures of helium and heavier ions, but at Jupiter and Saturn the proportion of heavy ions is much greater. These heavy ions are injected from planetary satellites in the magnetosphere (for example, sulfur and oxygen from Jupiter's volcanic moon Io). (Earth's moon is generally outside the magnetosphere.) At Uranus and Neptune the radiation belts are dominated by protons to a much greater extent than even at Earth. This is true despite the presence of numerous planetary satellites within these magnetospheres. The planetary magnetospheres and their radiation belts have intriguing similarities and differences that are presently understood only in part. More detailed comparison of planetary magnetospheres should yield important new insights into the processes that govern the space environment of our own planet Earth.

References

Gehrels, T. (Ed.), *Jupiter*, University of Arizona Press, Tucson, 1976.

Gehrels, T., and M. S. Matthews (Eds.), *Saturn*, University of Arizona Press, Tucson, 1984.

Hess, W. N., *The Radiation Belt and Magnetosphere*, Blaisdell Publishing Company, Waltham, MA, 1968.

Northrop, T. G., *The Adiabatic Motion of Charged Particles*, Interscience Publishers, New York, 1963.

Roederer, J. G., *Dynamics of Geomagneticlly Trapped Radiation*, Springer-Verlag, Berlin, 1970.

Schulz, M., and L. J. Lanzerotti, *Particle Diffusion in the Radiation Belts*, Springer-Verlag, New York, 1974.

Van Allen, J. A., *Origins of Magnetospheric Physics*, Smithsonian Institution Press, Washington, D.C., 1983.

James A. Van Allen

Department of Physics and Astronomy, University of Iowa, Iowa City, IA 52242.

Plasma Waves and Instabilities

S. Peter Gary

Just as a heated solid will melt into a liquid and a liquid heated further will evaporate into a gas, so will a gas subjected to yet further thermal input become ionized. In other words, if the atoms or molecules of a gas are given sufficient thermal energy, particle collisions can tear individual electrons away from these neutral particles so that the gas becomes a plasma, or a collection of negatively charged electrons and positively charged ions. Thus, the hot, dense fluid of the Sun's atmosphere [*Foukal*, this vol.] becomes a plasma at sufficiently high altitudes, and that state is maintained as this atmosphere expands outward to form the solar wind [*Goldstein*, this vol.].

Ultraviolet radiation can also kick electrons out of their atomic or molecular orbits. If a planet has a neutral atmosphere, that atmosphere acts as shield against the ionizing effects of such radiation from the Sun. But at sufficiently high altitudes, the tenuous atmosphere is ionized by solar ultraviolet radiation to form the planetary ionosphere [*Richmond*, this vol.] and magnetosphere [*Cowley*, this vol.]. Thus, plasma is the primary constituent of both planetary magnetospheres and the interplanetary medium.

Charged particles interact with one another through their intrinsic electric fields and through the magnetic fields generated by their relative motion. Neutral particles, such as the molecules of a gas, interact by strong collisions with one another; they do not sense one another until they are very close and then suffer "hard" collisions that cause large changes in their velocities. Electric and magnetic fields correspond to much longer-range forces than the molecular forces characterizing neutral particle interactions; since the charged particles of a plasma can sense one another at much greater distances, their interactions correspond to much "softer" collisions in which velocities change relatively little.

Neutral particle collisions primarily correspond to the interaction of two particles at a time; the molecular forces are of such short range that it is very unlikely that three particles can instantaneously collide among one another. However, in most tenuous solar system plasmas, the long-range character of charged particle interactions implies that any one particle interacts with many other charged particles at the same time.

In a neutral gas, information is carried by sound waves, pressure oscillations that are propagated by collisions and travel through the gas with a speed similar to the average atomic or molecular speed. Similarly, it is the electric and magnetic field interactions that permit waves to propagate through a plasma. But the long-range nature of the charged particle interactions implies that waves in a plasma are different in character from the sound wave in a gas.

The fundamental oscillation of plasma electrons takes place at the electron plasma frequency,

$$v_{pc} = (4\pi n e^2 / m_e)^{1/2}$$

where n is the electron density, e is the electronic charge, and m_e is the mass of the electron [*Chen*, 1974]. If a uniform distribution of electrons is locally disturbed, the resulting electric fields will pull the electrons back toward the uniform state. But once the particles return to this condition, they have some velocity, so they overshoot and create another disturbance with electric fields that once again pull the electrons back toward the uniform state, *ad infinitum*. The electron plasma frequency is the characteristic frequency of this oscillation; the more massive ions cannot respond at the relatively high electron plasma frequency and are essentially stationary with respect to this oscillation. Light waves below the electron plasma frequency cannot propagate through a plasma because the electric fields of such a wave are canceled out by the electron motion they induce.

In addition to light waves, there are two other waves that can propagate in a magnetic-field-free plasma. Electron plasma waves, like light waves, also can travel only above the electron plasma frequency. But, unlike light waves, these waves have no fluctuating magnetic field and are intrinsic to the plasma. Another intrinsic plasma mode is the analog of the sound wave in a neutral gas, the ion acoustic wave. Because this mode is based on the oscillations of the ions, it can propagate at frequencies below the electron plasma frequency; like the sound speed, its wave speed is similar to the average or thermal plasma ion speed. If a background magnetic field is also present, as it almost always is in solar system plasmas, many additional waves may arise. A characteristic frequency of waves in magnetized plasmas is the electron cyclotron frequency W_{ce} ∫ eB/m_ec, where B is the background magnetic field, and c is the speed of light. We do not have the opportunity to discuss waves in a magnetized plasma in detail here.

Plasma waves can lose energy by a mechanism known as Landau damping or collisionless dissipation [*Nicholson*, 1983]. If a plasma wave travels with sufficient speed relative to the charged particles of a plasma, those particles will sense an oscillating field and, over time, will neither gain nor lose much energy with respect to the wave. However, if a significant number of charged particles move through the plasma with the same speed and in the same direction as a plasma wave, the particles will experience relatively constant fields, and there can be a strong exchange of energy between the fields and these particles. If the plasma particles are in a near-equilibrium state, the net exchange of energy will be from the wave fields to the particles and the wave will diminish in amplitude, that is, it will be damped. The ion acoustic wave is a good example of this damping; if the electron temperature T_e is much greater than the ion temperature T_i, the ion acoustic wave speed is much greater than the ion thermal speed, few ions experience the wave-particle interaction, and the wave may propagate. However, if $T_e \sim T_i$, the wave speed is much closer to the thermal ion speed, there is a strong wave-particle interaction, and the ion acoustic wave is heavily damped.

Many solar system plasma are not close to thermal equilibrium. Plasmas of the solar wind and planetary magnetospheres continually experience changing flows and magnetic fields due to shock waves, reconnection events [*Cowley*, this vol.], and other dynamic activity. As a consequence, space plasmas are often observed to be far from the isothermal, isotropic condition characteristic of equilibrium gases. For example, a gentle compression of the magnetic field will make the distribution of plasma particle velocities anisotropic with the temperatures perpendicular to the magnetic field become larger than those parallel to that field. As another example, a strong localized heating of plasma, as at a shock, will lead to the heated particles escaping from their source region and flowing through a previously undisturbed remote plasma, corresponding to a strongly anisotropic plasma. Virtually every disturbance of a plasma—a gradient in density, temperature, or magnetic field, a velocity or field shear, the flow of a current or heat flux—corresponds to a nonthermal plasma.

Clearly, plasma dynamics are substantially different from the dynamics of neutral gases. And yet these differences rest on the same fundamental principles that govern all of classical nonrelativistic physics: Newton's laws, Maxwell's equations, and the laws of thermodynamics. In particular, plasmas, like all other physical systems, are subject to the Second Law of Thermodynamics: Entropy increases. In a neutral fluid or gas, a local disturbance that might contain some useful information that the second law requires to be dissipated is, in fact, rapidly smoothed by close encounter particle collisions. However, in many plasmas, neither close encounter collisions nor Landau damping can provide effective dissipation; there must be another mechanism that can quickly enforce the second law.

In many space plasmas, that mechanism is the growth of relatively short-wavelength plasma instabilities and the subsequent enhancement of wave-particle interactions. If a plasma is nonthermal, it is said to possess "free energy." If there is a sufficient amount of this quantity, the consequences of wave-particle interactions are turned around. Thus, instead of particles gaining energy from fields so that fluctuations are damped, the particles give up energy to the fields so that the electric and magnetic fluctuations grow in time, and an instability develops. A simple example here is a current associated with a relative drift between the electrons and protons. If this drift is relatively modest, the ion acoustic wave remains damped, albeit less strongly than at zero relative drift. But if the relative velocity difference between the two species becomes significantly greater than the ion thermal speed, the plasma becomes sufficiently anisotropic to permit the ion acoustic wave to grow in time. Because the mode is driven unstable by the relative motion of the electrons against the ions, we use the term "electron/ion acoustic instability" to describe this mode.

Different kinds of free energy lead to the growth of different kinds of plasma instabilities. For example, an anisotropy due to two groups of electrons with different average velocities streaming through each other can give rise to the electron/electron plasma instability. And, just as it increases the number of plasma waves, so does the presence of a background magnetic field substantially increase the number of possible instabilities [*Melrose*, 1986; *Gary*, 1993].

As the field fluctuations of the instability grow, they produce stronger wave-particle interactions that act to change the momentum and energy of the particles. Although fluctuation growth can be described by linear theory, nonlinear theory must be used to describe how the particles respond to the growing waves and how the fluctuations reach their maximum, or saturation, energy (as they must under the mandate of the First Law of Thermodynamics). The changes induced by wave-particle interactions are, of course, such as to reduce the free energy and return the plasma toward stable, more nearly thermal conditions. In the example of the electron/ion acoustic instability, the primary consequence of wave-particle interactions is to reduce the electron/ion relative drift back toward values corresponding to wave stability.

Instabilities play a role in many different space plasmas. The flow of thermal energy away from the Sun corresponds to a free energy in solar wind electrons. There is good evidence that this leads to heat flux instabilities that act to limit this thermal flow. The strong plasma perturbations associated with the planetary bow shocks and solar wind shocks cause the ions to become anisotropic; the resulting instabilities not only reduce these anisotropies but also accelerate a few of the particles to high energies. The plasma and field gradients associated with certain configurations at the magnetopause and in the magnetotail give rise to instabilities that are thought to trigger magnetic reconnection processes at these locations.

As a final example, consider the anisotropic distributions of energetic particles trapped in the Earth's radiation belts [*Van Allen*, this vol.] that lead to plasma instabilities. The enhanced wave-particle interactions resulting from these growing fluctuations act to make these distributions more isotropic; however, this increases the number of energetic particles that can move downward along the geomagnetic field lines. As these particles reach the Earth's atmosphere, they collide with and excite neutral gas molecules, causing them to emit radiation that we recognize as auroral phenomena [*Akasofu*, this vol.]. In this way, plasma instabilities make important contributions to the aurora.

More generally, plasma instabilities are worthy of study because they offer the potential of relatively complete understanding through the use of existing spacecraft data. The small-scale, local nature of many plasma instabilities implies that present-day particle and field measurements from single spacecraft may be sufficient for theoreticians to develop comprehensive models for instability growth and the consequences of wave-particle interactions. The same cannot be said for studies of larger-scale space plasma phenomena, which will require future multiple-spacecraft missions to provide the data necessary for a full understanding of their properties.

Furthermore, the local nature of plasma instabilities implies broad applicability for successful theories of their development. For example, an understanding of whether and how instabilities contribute to reconnection in the terrestrial magnetosphere is very likely to improve our ability to explain the same process in the magnetospheres of Jupiter, Saturn, and other planets, including those not yet discovered. And the understanding we may gain from space plasma instabilities may have application to a variety of laboratory plasmas, including those used in fusion research.

Finally, although global modeling of space plasmas does not presently make much use of plasma instability information, that situation is likely to change as increased computing capability permits modelers of large-scale space plasma problems to address finer spatial scales and more detailed physical properties.

References

Chen, F.F., *Introduction to Plasma Physics*, Plenum Press, New York, 1974.

Gary, S.P., *Theory of Space Plasma Instabilities*, Cambridge University Press, Cambridge, 1993.

Melrose, D.B., *Instabilities in Space and Laboratory Plasmas*, Cambridge University Press, Cambridge, 1986.

Nicholson, D.R., *Introduction to Plasma Theory*, John Wiley and Sons, New York, 1983.

S. Peter Gary

Los Alamos National Laboratory, Mail Stop D438, Los Alamos, NM 87545.

The Ionosphere and Upper Atmosphere

A. D. Richmond

B ecause our society is becoming increasingly dependent on technological systems that can be affected by ionospheric phenomena during
geomagnetic storms, the ionosphere, its electrodynamics, and its coupling with the neutral atmosphere and the magnetosphere are being studied as part of a coordinated program of "space weather" research. This
research seeks to characterize the variability of ionospheric density and
electric currents during magnetic storms, and to determine to what extent
valid predictions of those phenomena and their effects can be made.

The Ionosphere

The upper atmosphere contains free electrons and ions produced by
ionizing radiation from the Sun and from the Earth's space environment. It
comprises a weakly ionized plasma, called the ionosphere, that conducts
electricity. Above 60-km altitude electrons are sufficiently dense to influence the propagation of radio waves, giving the ionosphere much of its
practical importance. The ionosphere lies at the base of the magnetosphere,
which encompasses those regions of space where the Earth's magnetic field
has a dominant influence on charged particles. The electrodynamical
behavior of the ionosphere is strongly influenced both by the neutral
atmosphere it is embedded in and by the magnetosphere. Global electric
currents flow throughout the ionosphere and magnetosphere, connecting
into currents in interplanetary space that are carried by the plasma of the
solar wind. The highly variable currents and their associated electric fields
have a major impact on the energetics, dynamics, and structure of the
upper atmosphere and the space environment.

The ionosphere's influence on radio wave propagation is sometimes useful, but sometimes bothersome. Over-the-horizon transmissions for telecommunications or surveillance usually rely on ionospheric reflection at radio frequencies below about 30 MHz. The maximum usable frequency depends, among other things, on the maximum electron density of the ionosphere, which is highly variable. Lower-frequency waves are subject to absorption in the lower ionosphere, where the electrons oscillating in the wave's electromagnetic field lose energy to air molecules through collisions. The radio wave absorption depends on the electron density, and is strongest during the day. The absorption can increase to the point of radio blackout during sporadic ionization enhancements associated with solar flare X rays [*Lin*, this vol.] and with energetic protons that precipitate into the high-latitude upper atmosphere [*Tsurutani and Gonzalez*, this vol.].

Because the ionosphere is a magnetized plasma, it is important in the study of plasma physics. A variety of natural plasma instabilities occur that are observed with radars and other radio wave techniques, as well as with rockets and spacecraft. Active experiments are carried out by modifying ionospheric properties with high-power radio waves, with chemical releases, or with space-based energetic electron beams. Unlike plasmas studied in the laboratory, the ionosphere has no chamber walls to interfere with the experiments or to complicate interpretation of the data.

Radio transmissions between the Earth and spacecraft operate at frequencies that are not reflected by the ionosphere and that do not suffer much absorption. However, these transmissions are subject to degrading scintillation when they refract through ionospheric irregularities. They also undergo phase-path changes and propagation delays in traversing to ionosphere, necessitating adjustments to precise measurements like satellite-based radar altimetry of ocean and land surfaces, positioning with the Global Positioning Satellite (GPS) system, and radio astronomy. To some extent, the electron density and the irregularities vary predictably with the time of day, season phase of the 11-year sunspot cycle, and geographic location. However, they are also subject to irregular variations due to influences from the magnetosphere and lower atmosphere.

Ionospheric electric currents, especially those strong currents that occur during magnetic storms, can have a number of impacts. The magnetic field produced by the currents induces additional electrical currents in the Earth that can flow through grounded electrical power grids and harm their transformers or trip circuit breakers [*Joselyn*, this vol.]. On occasion, magnetic storms cause large-scale disruptions of power grids, as happened in Quebec on March 13, 1989. Even during less disturbed periods, the magnetic perturbations associated with ionospheric currents complicate geomagnetic surveys that attempt to derive accurate models of the Earth's internal field or study subtle spatial structure in the field. Electrical heating of the upper atmosphere above 120 km during storms raises the temperature, thereby reducing the rate of exponential density falloff with increasing altitude so that the density at high altitude is greatly increased. Satellites orbiting the Earth below 1000 km then experience perceptible alterations of their trajectories owing to the increased atmospheric drag. They can become temporarily lost to satellite-tracking services. The heating also changes the wind patterns and the composition of the upper atmosphere, which influence the plasma density distribution.

The Conducting Upper Atmosphere

The earliest suggestions that the rarefied upper atmosphere might conduct electricity came from eighteenth-century experimenters who were struck by the similarity of polar lights (auroras) and electrical glows produced in evacuated containers. Benjamin Franklin formulated a theory of the aurora that invoked accumulations of electrical charge at polar latitudes. Experimenters of that time also noticed fluctuations of magnetic compass needles, which led 19th-century scientists like Carl Friedrich Gauss to suggest that electric currents might flow in the upper atmosphere. Around 1880, Balfour Stewart proposed that the currents could be driven by upper atmospheric winds that would generate electromotive forces in the conducting medium as they move it through the geomagnetic field, in effect acting as an electric dynamo. In 1908, Kristian Birkeland suggested that strong currents in the auroral ionosphere during magnetic disturbances are caused by charged particles from distant space that are forced to flow primarily along geomagnetic field lines until they reach the high-latitude ionosphere. Marconi's demonstration in 1901 that radio waves could propagate across the Atlantic led to suggestions by Kennelly and Heaviside that the waves might be reflected by a conducting layer in the upper atmosphere. In 1924 experiments by Appleton and Barnett and by Breit and Tuve demonstrated that direct reflection of radio waves from about 100 km altitude, thus clearly establishing the existence of the ionosphere.

The main ionosphere is generally divided into D, E, and F regions, at altitudes of roughly 60–90 km, 90–140 km, and 140–1000 km, respectively, based on features of the electron-density profile with altitude (Figure 1). The primary ionization sources are solar ultraviolet and X-ray radiation at wavelengths shorter than 103 nm striking the day side of the Earth, and energetic electrons precipitating from the magnetosphere into the auroral regions. Solar ultraviolet and X-ray radiation vary over the 11-year solar activity cycle [*Foukal*, this vol.], with sporadic enhancements during solar flares. Precipitating auroral electrons vary dynamically in association with magnetospheric disturbances.

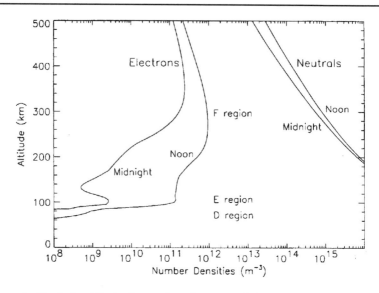

Figure 1. Typical midlatitude number densities of electrons and of neutral molecules at noon and midnight. The positive-ion number density is essentially identi-
cal to the electron density in the E and F regions, but is greater than the electron
density in the D region, where negatively charged ions are also present.

The ionospheric electron density is highly variable, depending not only
on the ionization sources, but also on ion-neutral chemical transformations,
ion-electron recombination, and plasma transport by neutral winds, electric
fields, and diffusion. The maximum density in a vertical profile usually
occurs in the F region between about 200 and 500 km, with values between
4×10^{10} m^{-3} and 4×10^{12} m^{-3}, which correspond to natural resonant plasma
frequencies of 1.8–18 MHz. Radio waves are totally reflected at frequencies
below the maximum resonant frequency, called the critical frequency.
Reflection can also occur at higher frequencies for waves obliquely incident
on the ionosphere. Obviously, long-distance terrestrial radio transmissions
that reflect from the ionosphere must rely on frequencies below or near the
critical frequency, while radio astronomy and communications with spacecraft can only use those frequencies that penetrate the ionosphere.
    The ionospheric electrical conductivity is highly anisotropic, owing to
the strong influence of the geomagnetic field on charged-particle motion.
At high altitudes, where collisions between ions and neutral air molecules
are infrequent, the ions and electrons gyrate around magnetic-field lines,
though they are free to move parallel to the field, the direct-current conductivity along the magnetic field can be as large as 100 S/m, while the conductivity perpendicular to the field is usually less than 10^{-4} S/m above 150

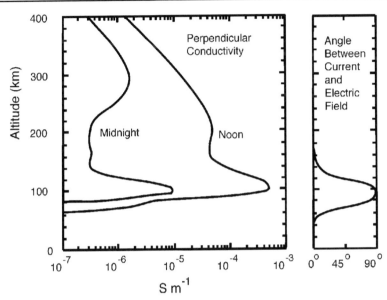

Figure 2. (Left) Typical magnitudes of the conductivity perpendicular to the geomagnetic field at noon and midnight for midlatitudes. (Right) The angle between the directions of the current and the electric field (in the frame of reference moving with the conducting medium), both in the plane perpendicular to the geomagnetic field. As viewed along the magnetic field, the direction of the perpendicular current is rotated clockwise from that of the electric field. The conductivity along the magnetic field (not shown) is much larger than the perpendicular conductivity at all heights above 90 km.

km. (For comparison, seawater has a conductivity of about 4 S/m.) The large parallel conductivity almost completely shorts out any electric field that might otherwise be established along the magnetic field, so that for most situations the geomagnetic field lines can be considered to be electric equipotentials. At lower altitudes, collisions between the ions and neutrals become more frequent, decoupling the electron and ion motions in the plane perpendicular to the magnetic field, so that more significant amounts of current can flow in that plane. The direct-current conductivity perpendicular to the magnetic field is largest at heights of 90–150 km during the day and in the nighttime auroral zone, with maximum values of the order 10^{-3} S/m (Figure 2). In addition to the parallel/perpendicular anisotropy of the conductivity, further anisotropy occurs perpendicular to the geomagnetic field: The Hall effect causes the direction of the current to deviate by as much as 88° from that of the electric field, an effect maximizing around 100 km altitude. For alternating currents at radio frequencies, the conductivity is frequency-dependent. I becomes nearly isotropic at frequencies well above the gyrofrequency of electrons in the geomagnetic field (on the order of 1 MHz), and it decreases with increasing frequency.

Sources of Ionospheric Electric Fields and Currents

The two main sources of global-scale electric-field generation in the ionosphere are the ionospheric wind dynamo and the solar-wind/magnetospheric dynamo. A third source, thunderstorm activity, is believed to contribute only in a minor way to the global-scale fields, though at night, it may be locally important.

The ionospheric dynamo is essentially that mechanism proposed by Stewart: Winds in the thermosphere (90–500 km) move the conducting medium through the geomagnetic field, producing an electromotive force (emf) that drives currents and sets up polarization electric fields. Electric-potential differences of 5 to 10 kV between different parts of the globe are produced by this mechanism. The emf interacts only with the conductivity component transverse to the geomagnetic field, so that dynamo action is weighted toward the 90–150 km height range during the day. At night, however, the E region transverse conductivity is greatly diminished, so that F region dynamo action above 200 km becomes more important. The ionospheric currents are strongest on the dayside of the Earth, where they typically form two large horizontal current vortices, clockwise in the southern hemisphere and counter-clockwise in the northern hemisphere. The currents in the two hemispheres are connected by a magnetic field-aligned current when the dynamo effects in the two hemispheres are unbalanced.

The solar wind/magnetospheric dynamo draws its energy from the kinetic and thermal energy of the solar wind and magnetospheric plasmas, generating electric fields and currents that connect to the high-latitude auroral and polar ionosphere along geomagnetic field lines, as suggested by Birkeland [*Cowley*, this vol.]. This interaction depends strongly on the direction of the interplanetary magnetic field that is embedded in the solar wind, since the direction of that field determines the topology of its connection with the Earth's magnetic field. The ionospheric electric fields and currents produced by the solar wind/magnetospheric dynamo are usually much stronger than those of the ionospheric wind dynamo, and are highly variable in time. On average, a high electric potential is established around 70–75° magnetic latitude on the morning side of the Earth, and a low potential at around the same latitude on the evening side. The potential drop varies from 20 to 200 kV.

Electric power generation by the dynamos involves extraction of energy from the thermospheric wind and from the solar wind and magnetospheric plasmas, modifying these in the process. For example, thermospheric winds experience a significant drag force as the electric currents they generate flow through the geomagnetic field. This force is known as "ion drag" because it results from collisions between ions and neutral molecules moving at different mean velocities. The solar wind is also retarded by the dynamo currents it generates. These energy losses, as well as the rates of electric energy transfer between the ionosphere and the hot magnetospheric plasma, are dependent on the ionospheric conductivity. The ionospheric conductivity is itself dependent on the dynamo electric fields, since those fields cause transport and redistribution of the F region plasma. Furthermore, the electrical circuits of the ionospheric wind dynamo and of the solar wind/magnetospheric dynamo are intercoupled, so that the two dynamos react to each other. For example, the strong high-latitude currents driven by the solar wind/magnetospheric dynamo, flowing through the geomagnetic field, force high-speed thermospheric winds by a motor effect, winds that in turn influence the electric fields and currents. Realistic modeling of dynamo processes quickly becomes very complicated when the various feedback effects are considered. This is an active area of current research in ionospheric electrodynamics.

The electric power is used up in a number of ways. Much of it is dissipated as resistive heating in the thermosphere, especially at high latitudes. Some of it is transferred through the ionospheric circuit from the solar wind source to the magnetospheric plasma, which is heated as it is transported toward the Earth into regions of stronger magnetic field. Some of it goes into acceleration of energetic electrons that precipitate into the high-latitude thermosphere to produce the polar lights. A fraction of the electric power goes into forcing the strong high-latitude thermospheric winds.

Inferences from Observations of Dynamo Effects

Observations of ionospheric electric fields and of the magnetic perturbations produced by ionospheric currents give us important information about thermospheric winds and the interaction of the solar wind with the magnetosphere. Direct observations of thermospheric winds are relatively limited, but observations of magnetic perturbations exist for long periods of time at many locations around the Earth. When interpreted with the aid of simulation models of the ionospheric wind dynamo, magnetic data from sites at middle and low latitudes can provide a wealth of information about the distribution and variability of thermospheric winds on the sunlit side of the Earth. At high magnetic latitudes, observations of ionospheric electric fields and of magnetic perturbations on the ground and on satellites reveal characteristics of solar wind/magnetospheric dynamo processes.

At middle and low latitudes, winds in the ionospheric dynamo region tend to be dominated by global oscillations. Above 140 km, daily wind oscillations with magnitudes over 100 m/s are driven primarily by the absorption of far-ultraviolet solar radiation. Between 90 and 140 km the oscillations are strongly influenced by upward propagating global waves, called atmospheric tides, that are generated by solar heating at lower altitudes: in the upper ozone layer and in the troposphere. Gravitational tidal forcing by the Moon and Sun also contribute, but only minimally. As the tides propagate into regions of exponentially decreasing air density, their amplitudes can grow, reaching values of 100 m/s or so in the lower thermosphere before the waves dissipate. The generation and propagation conditions for these waves tend to favor the arrival of semidiurnal (12-hour) tides over diurnal (24-hour) tides in the dynamo region. Upward propagating planetary waves with periods of 2–20 days are also believed to influence winds in the lower thermosphere, but their relative importance there has not yet been established. At high latitudes, electric currents drive thermospheric winds that at times can reach 1000 m/s or more in the upper thermosphere, both by the motor effect mentioned earlier and by resistive heating of the gas that affects the pressure gradient forces on the air. Variations in the sources of the winds, as well as variations in the propagation conditions of tides and planetary waves through the middle atmosphere, are responsible for variability of the thermospheric winds. Analyses of geomagnetic variations have revealed many properties of the winds and their variations. Since many of the geomagnetic measurements extend back for many decades, studies related to possible long-term global atmospheric change are feasible.

The high-latitude ionosphere provides a window to the outer magnetosphere, since electric fields, electric currents, and energetic charged particle populations in the magnetosphere readily project along magnetic field lines down to the ionosphere, where they are more easily measured. Throughout much of space, the electric field that would be measured in the frame of reference moving with the plasma nearly vanishes, because the charged particles rapidly adjust to cancel an electric field. If E and v are the electric field in the frame moving with the plasma is $E + v \times B$ (for a nonrelativistic Lorentz transformation). Then

$$E + v \times B \sim 0 \qquad (1)$$

meaning that the electric field and the plasma velocity are closely linked. As electric fields project along the magnetic field, so do plasma motions.

Figure 3 shows an example of electric potential patterns in the northern and southern polar regions deduced from syntheses of data from many dif-

Northern Hemisphere $B_z < 0$, $|B_y| > |B_z|$ Southern Hemisphere

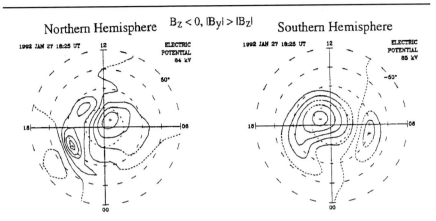

Figure 3. Polar ionospheric electric-potential patterns in the northern (left) and southern (right) hemispheres for January 27, 1992, 1825 UT, when the interplanetary magnetic field had components southward and dawnward [after *Lu et al.*, 1994]. For both hemispheres the ionosphere is viewed from above the north magnetic pole; thus for the southern hemisphere the ionosphere is viewed as though one were looking through the Earth. The coordinates are magnetic local time, increasing counterclockwise from midnight at the bottom, and magnetic latitude, poleward of 50°. The contour interval is 10 kV. Dashed contours indicate regions of large uncertainty in the estimated potentials.

ferent ground- and satellite-based instruments. Both hemispheres are viewed from above the north magnetic pole (the southern hemisphere is thus viewed through the Earth), so that for both hemispheres the geomagnetic field is directed into the page. Equation (1) implies that plasma convects along electric potential contours, counterclockwise around the potential highs (+) and clockwise around the lows (–). The interplanetary magnetic field at this time has a dusk-to-dawn component. In the northern hemisphere, magnetic field lines emanating from the Earth (outward) above about 78° on the dayside eventually bend to the left at great distance to join the interplanetary field, while in the southern hemisphere magnetic field lines emanating from the Earth (inward) above about 78° on the dayside eventually bend to the right. Magnetic field tension tends to pull the plasma toward dusk in this region in the northern hemisphere, and toward dawn in the southern hemisphere, while the solar wind also drags the plasma on these field lines across the pole toward midnight. At lower latitudes the geomagnetic field lines are no longer connected to the interplanetary field, and the plasma flow can return toward the dayside. When the interplanetary field changes direction, as it frequently does, the plasma convection in the polar ionospheres also changes. Currently, there is intensive research into understanding magnetospheric processes at the boundaries between geomagnetic field lines that interconnect with the interplanetary field and those that do not, corresponding to roughly 80° magnetic latitude in the dayside ionosphere and 70° on the nightside. At these boundaries the approximation of equation (1) breaks down, and plasma processes become much more complex.

Acknowledgments. I thank Steve Suess, Bruce Tsurutani, and Robert Stening for helpful comments. This work was supported by the NASA Space Physics Theory Program and by NASA Order W-17,385. The National Center for Atmospheric Research is sponsored by the National Science Foundation.

Additional Reading

Hargreaves, J. K., *The Solar-Terrestrial Environment: An Introduction to Geospace—The Science of the Terrestrial Upper Atmosphere, Ionosphere, and Magnetosphere,* Cambridge University Press, New York, 1992.

Kelley, M. C., *The Earth's Ionosphere: Plasma Physics and Electrodynamics,* Academic Press, San Diego, CA, 1989.

Kivelson, M. G., and C. T. Russel (Eds.), *Introduction to Space Physics,* Cambridge University Press, New York, 1995.

Lu, G., et al., Interhemispheric asymmetry of the high-latitude ionospheric convection pattern, *J. Geophys. Res., 99,* 6491, 1994.

Rees, M. H., *Physics and Chemistry of the Upper Atmosphere,* Cambridge University Press, New York, 1989.

Richmond, A. D., The ionospheric wind dynamo: Effects of its coupling with different atmospheric regions, in *The Upper Mesosphere and Lower Thermosphere,* edited by R. M. Johnson and T. L. Killeen, pp. 49–65, AGU, Washington, D.C., 1995.

Volland, H., *Atmospheric Tidal and Planetary Waves,* Kluwer Academic Publishers, Dordrecht, The Netherlands, 1988.

Volland, H. (Ed.), *Handbook of Atmospheric Electrodynamics,* vol. II, CRC Press, Boca Raton, FL, 1995.

A. D. Richmond
High Altitude Observatory, National Center for Atmospheric Research, Boulder, Colorado

Red Sprites and Blue Jets: Transient Electrical Effects of Thunderstorms on the Middle and Upper Atmospheres

D. D. Sentman and E. M. Wescott

Four new and diverse classes of energetic electrical effects of thunderstorms have been documented over the past 5 years. Two of these classes, called red sprites and blue jets, are large-scale optical emissions excited by lightning. Together they span the entire distance between tops of some thunderstorms and the ionosphere. Gamma-ray (1 MeV) bursts and extremely intense VHF radio bursts some 10^4 times larger than normally produced by lightning have been observed from low Earth orbit and are also believed to originate in thunderstorms. Taken together, these newly discovered classes of natural electrical phenomena provide evidence that thunderstorms are both more energetic and capable of electrically interacting with the upper atmosphere and ionosphere to a far greater degree than has been appreciated in the past. Here, characteristics of red sprites and blue jets are summarized.

1. New High-Altitude Electrical Phenomena

The serendipitous and remarkable low-light-level television observation by *Franz et al.* [1990] of large-scale optical emissions high above a midwestern thunderstorm paved the way for the discovery of several new classes of previously unrecorded high-altitude atmospheric electrical phenomena. Since this initial report, well over a thousand images of similarly brief luminous structures have been observed in the mesospheric D-region above thunderstorms using low-light-level video systems. Most of these images have been obtained from the ground [*Lyons*, 1994; *Lyons and Williams*, 1994; *Winckler*, 1995] and from aircraft [*Sentman et al.*, 1993, 1995], but about 20 events have also been observed from the space shuttle above thunderstorms on the limb of the planet [*Boeck et al.*, 1994]. Early reports referred to these events by a variety of names, but now they are simply called "sprites," a term that is succinct and whimsically evocative of their fleeting nature while avoiding making unwarranted implications about as yet unknown physical processes.

A second, previously unrecorded, and equally remarkable form of electrical activity above thunderstorms has also been reported recently. "Blue jets" [*Wescott et al.*, 1995] are sporadic optical emissions that erupt in narrow cones directly from the tops of clouds and shoot upward through the stratosphere. Blue jets appear to be a class of phenomena distinct from sprites.

In addition to sprites and blue jets, two other new types of unexpected emissions have been detected that also appear to originate in thunderstorms. Short duration (~1 ms) gamma-ray (>1 MeV) bursts of terrestrial origin have been detected by the Compton Gamma Ray Observatory [*Fishman et al.*, 1994]. Their source is believed to lie at altitudes greater than 30 km somewhere above the thunderstorm. Finally, extremely intense pairs of VHF pulses (Trans-Ionospheric Pulse Pairs (TIPPS)) originating from thunderstorm regions, but some 10^4 times stronger than VHF sferics produced by normal lightning activity, have been observed by the ALEXIS satellite [*Holden et al.*, 1995].

The possible existence of sprites and blue jets was suspected through scattered anecdotal visual reports dating back more than a century, but they remained unrecorded until little more than 5 years ago. Gamma-ray bursts and TIPPS were discovered less than 2 years ago. The element that is common to all of these fascinating events is the thunderstorm, but beyond this, very little is known about the details of the physical processes associated with these new classes of atmospheric electrical processes.

In this article we summarize the basic observational features of the entries in this small but expanding catalog of newly discovered atmospheric electrical processes that are visible, the sprites and blue jets.

2. Middle and Upper Atmospheric Flashes

Red Sprites

Sprites are very large luminous flashes that appear within the mesospheric D-region directly over active thunderstorm systems coincident with cloud-to-ground or intracloud lighting strokes. To the unaided human eye they are brief and only barely detectable, but in intensity-enhanced television images sprites appear in a dazzlingly complex variety of forms. Their spatial structures range from small single or multiple vertically elongated spots, to spots with faint extrusions above and below, to bright groupings containing dozens of separated elements. Triangulation of their locations and physical dimensions using simultaneous images captured from widely spaced aircraft has shown that their terminal altitude extends to the ionosphere. The brightest region of a sprite is red and lies in the altitude range 65–75 km. Above this there is often a faint red glow or wispy structure extending upward to about 90 km, to the nighttime E-region ledge. Below the bright red region, blue tendril-like filamentary structures often extend downward to as low as 40 km (see Figures 1 and 2).

Sprites rarely appear singly, usually occurring in clusters of two, three, or more. Some of the very large events, such as those shown in Figure 1, seem to be tightly packed clusters of many individual sprites. Other events consist of loosely scattered elements extending across horizontal distances of 40 km or more. Large sprite clusters may occupy volumes in excess of 10^4 km^3 in the middle atmosphere.

High-speed photometer measurements show that individual elements of sprite clusters appear suddenly, coincident with cloud lightning below, and persist for no more than a few milliseconds. Large sprite clusters formed by the appearance of individual sprite elements in rapid succession occasionally give the visual impression of "dancing" horizontally across the sky above the thunderstorm.

Current evidence strongly suggests that sprites preferentially occur above decaying regions of mesoscale thunderstorms in coincidence with large positive cloud-to-ground lightning strokes. Their occurrence rate appears to be approximately one for every 100-200 negative strokes. The optical intensity of sprite clusters, estimated by comparison with tabulated stellar intensities, is comparable to a moderately bright auroral arc. The optical energy is roughly 10–50 kJ per event, with a corresponding optical power of 5–25 MW. Assuming that optical energy constitutes 10^{-3} of the total for the event, the total energy and power are on the order of 10–100 MJ and 5–50 GW, respectively.

Blue Jets

Blue jets are a second class of high-altitude optical emission, reported on rare occasions by pilots, that have only recently been recorded using low-light-level television systems. Confirmation of the existence of these events was first obtained unexpectedly from two jet aircraft circling an intense thunderstorm over Arkansas during a NASA-sponsored mission to study sprites. Numerous video images of the blue jets were captured simultaneously aboard both aircraft while peering across the anvil top from an altitude of about 40,000 feet toward the convective overshoot of the electrically active hot tower of the storm.

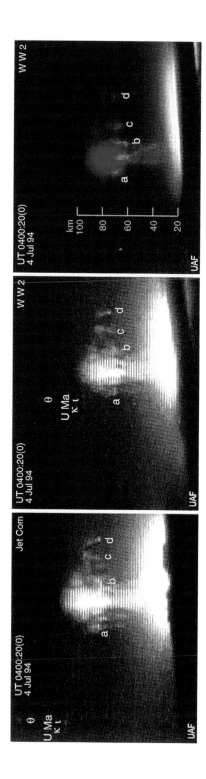

Figure 1. A large sprite event simultaneously observed from two NASA research jet aircraft on the night of July 4, 1994. The two monochrome images show the same event viewed from about 50 km apart. The parallax of the event as viewed from the two aircraft can easily be seen by comparison with two stars in Ursa Major. The righthand image shows the event as observed using the low-light-level color camera aboard one of the aircraft. The altitude scale has been determined by triangulation of the image with respect to the stars.

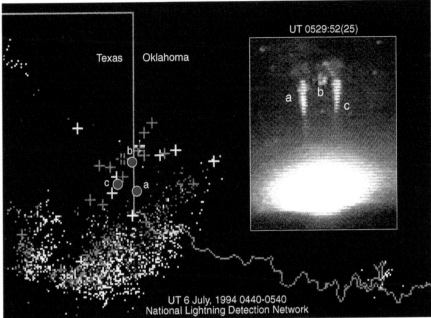

Figure 2. Spatial location of individual sprites in an event observed on the Texas-Oklahoma border the night of July 5, 1994 (UT July 6). The locations on the ground below the sprites are projected onto a time/polarity color-coded map of cloud-to-ground lightning. The sprites are seen to be laterally distributed over distances comparable to the terminal heights of the events (~100 km) and to occur in regions dominated by positive cloud-to-ground strokes.

As their name implies, the jets are sporadic optical ejections, deep blue in color, that appear to erupt from the vicinity of the overshoot. Following their emergence from the tops of the thundercloud, blue jets propagate upward in narrow cones of about 15 degrees full width at vertical speeds of roughly 100 km/s (Mach 300), fanning out and disappearing at heights of about 40-50 km over a lifetime of about 300 ms (see Figure 3). Their intensities are on the order of 600 kR near the base, decreasing to about 10 kR near the upper terminus. These intensities correspond to an estimated optical energy of about 4 kJ, a total energy of about 30 MJ, and an energy density on the order of a few mJ/m^3. Blue jets are not aligned with the local magnetic field.

The first image of both sprite and jets were obtained accidentally in the course of studying other phenomena and came as unexpected surprises. But they can be seen with the naked eye under the right viewing conditions, and they appear to be not uncommon features of many thunderstorms systems. Why, then have they eluded documentation for so long? Several factors may be responsible:

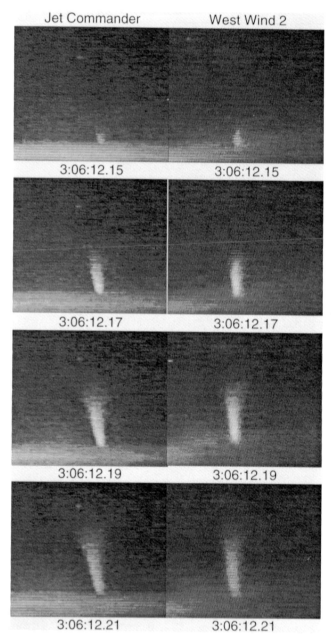

Figure 3. Stereo time sequence of a blue jet observed above a thunderstorm in southern Arkansas on the night of July 4, 1994. The images come from two GPS-synchronized low-light-level cameras aboard two aircraft separated by about 50 km. The jet is seen to erupt from the top of the cloud and propagate upward at a speed of 100 km/s to a terminal altitude of about 45 km.

(1) Sprites and blue jets occur high *above* active thunderstorm systems, a region often obscured from view from the ground by clouds. To observe them requires unobstructed visual access to the region from the storm, viewing against a dark stellar background.

(2) Sprites and blue jets are visible to the unaided eye only when it is dark-adapted. Their average brightness is roughly that of moderately bright aurora. In the human eye, this corresponds approximately to the crossover threshold intensities of cones on the retina, which respond to color, and the somewhat more sensitive but achromatic parfoveal rods, which permit night vision. The dark-adapted eye most readily sees sprites in parfoveal vision, when not directly looking at them.

(3) Sprites appear to have a duration of only a few (3–10) ms, too brief to shift one's gaze to obtain a visual fix in the event they are detected outside of the center of vision. Blue jets persist long enough to obtain a visual fix, but their occurrence seems to be restricted to small regions directly above the electrically active core of a thunderstorm. This region is often visually accessible from aircraft, but not from the ground.

(4) Cloud illumination from sprite- and jet-producing cloud-to-ground or intracloud lightning activity is often orders of magnitude brighter than sprites. This lightning activity can easily distract the casual observer, whether on the ground, in the air, or in space, from noticing the occasional blue jet or the fleeting and delicate dance of red sprites in the night sky high above the storm.

Given these factors, all of which conspire to make unaided visual detection of sprites and jets difficult, it is perhaps not surprising in retrospect that they have gone unrecorded until recently. Similar considerations apparently hold for their detection by the vast array of modern instruments monitoring the natural environment in virtually all energy bands ranging from below radio to above optical frequencies, both from the ground and from space.

4. Mechanisms

Theories proposed to date to account for red sprites can be classified into five categories (see the review by *Sentman and Wescott* [1995], and references therein) with some overlap between categories. All involve lightning discharges acting either as a causative agent, or as a simultaneous but non-causative consequence of electrical breakdown triggered by cosmic rays. They may be briefly listed as follows:

(1) Fluorescence/luminescence of middle or upper atmospheric gases. In this scenario soft x-ray or ultraviolet emissions from lighting occurring near cloud tops photoexcite molecular vibrational states. Subsequent decay to the ground state involves emission of an optical photon, producing the

observed sprites. Quasistatic or electromagnetic fields produced by the thunderstorm electric field or lightning do not play a role in this model.

(2) Quasistatic dielectric breakdown between electrified clouds and the high atmosphere. A large-scale electrostatic field is impulsively created in the middle and upper atmospheres following a large positive cloud-to-ground stroke. The enhanced electrostatic field above the cloud breaks down the ambient air, producing sprites and air glow.

(3) Quasistatic heating and impact ionization of the middle atmosphere. This mechanism is similar to (2), except that optical emissions from N_2 occur following impact ionization by electrons accelerated in the impulsively enhanced electrostatic field above a thunderstorm created by cloud-to-ground lightning stroke. Breakdown is not required to produce the observed emissions.

(4) Cosmic ray triggered runaway breakdown. Electrical breakdown of the high atmosphere from a cascade of runaway electrons is initiated by a cosmic-ray shower into the pre-existing region of intense electric field above an electrified cloud. The breakdown leads to the observed optical emissions.

(5) Radio frequency breakdown by lighting electromagnetic pulse. A strong transverse electromagnetic wave pulse propagates upward from intra-cloud or cloud-to-ground lightning discharges. The electric field in the pulse is strong enough to break down the air locally as it propagates through the middle and upper atmosphere, producing the observed emissions.

Whereas numerous physical mechanisms have been proposed to account for red sprites, to date (June 1995) no mechanisms have been advanced to explain blue jets.

5. Current Research

Intensive efforts, both experimental and theoretical, are underway to determine the physical mechanisms at work to produce red sprites, blue jets, gamma-ray bursts, and TIPPS. Initial thought is being given to possible roles these events, or processes that produce them, may play in the larger terrestrial electrical environment. Although optical images seem likely to remain an important form of experimental "ground truth," focus has already shifted to employing other diagnostics that will yield more specific information about detailed physical mechanisms. These include high-speed videography, optical and infrared spectra, radio (ELF-HF) measurements of the electromagnetic emissions from sprites and their accompanying tropospheric lightning strokes, VLF measurements of associated ionospheric heating effects and Trimpi events, balloon-borne in situ measurements of electric and magnetic fields directly above thunderstorms, and continuous wave radar measurements to determine electron densities.

6. Concluding Remarks

Solar-terrestrial physics is concerned with the elucidation of the flow of particles and fields from the Sun into the terrestrial system and their subsequent distribution and dissipation. It is not yet known whether red sprites, blue jets, gamma-ray bursts, TIPPS, or their underlying processes are important terminal links in the solar-terrestrial chain extending from the surface of the Sun, through the solar wind, across the magnetosheath and magnetosphere into the ionosphere, mesosphere, and lower atmosphere.

At first glance it would seem that these events are not important links in the usual sense, since the globally averaged energy densities are comparatively small and the energy flow is in the wrong direction, i.e., from sources within thunderstorms upward to sinks in the high atmosphere, ionosphere, and magnetosphere. It is unclear whether the absorption within the ionosphere or magnetosphere of this energy flowing upward from the lower atmosphere is capable of producing effects that are dynamically significant enough to qualify as a strong link between these elements. However, several long-lived secondary or tertiary effects within the neutral upper atmosphere may occur by way of joule heating, photoexcitation, or electron impact excitation or ionization of the air. For example, ionization or electronic excitation, by RF electrolysis or other means, could conceivable lead to the creation of reactive species or to the activation of catalytic species that would otherwise be absent. Likewise, the sudden appearance of several thousand cubic kilometers of joule-heated gas within a sprite should create an acoustic pulse with a characteristic period of about 100 seconds. Depending on its total energy and the amount that couples into a gravity wave, this infrasonic pulse could conceivably provide a significant contribution to vertical mixing in the middle atmosphere.

From what is known to date, it may be speculated that sprites or jets, or both, are an integral feature of most mesoscale thunderstorm systems of moderate (~100-km diameter) or larger size, and may be an essential element of the Earth's global electrical circuit. There is no reason to suppose they have not been regular features in the sky above thunderstorms for millions of years or longer. One may even speculate that similar phenomena should also occur on other plants with lightning, most notably, Jupiter and Venus.

Understanding where these new electrical processes fit in the terrestrial system is a challenging, and inherently multidisciplinary, problem that spans traditional discipline boundaries separating the lower and upper atmospheres. It will require researchers to consider thunderstorm dynamics, charge separation mechanisms in the troposphere, quisistatic and electromagnetic pulse effects on the stratosphere and mesosphere, the plasma- and photo-chemistry of the mesosphere, impulsive and bulk heating effects within the mesosphere and lower ionosphere, and their interaction with the magnetosphere and subsequent back-reaction on the atmosphere.

Acknowledgments. Besides the authors, the Sprites and Jets Research Team at the University of Alaska includes Daniel Osborne, Don Hampton, Matt Heavner, and Jim Desrochers, all of those dedicated efforts have contributed to the UAF research program. We also thank Hans Nielsen, Tom Hallinan, Neal Brown, and the Alaska Color Television Project for generously making available equipment for field observations. We would like to acknowledge Aero Air, Inc., of Hillsboro, Oregon, with whom we have had a very fruitful research association, and Rick Howard of NASA Headquarters. Finally, because of space limitations we have limited most of our discussion and references to selected primary video observations. There is a growing body of related observations and theoretical work, not covered here in detail, being carried out at Mission Research, Inc., MIT, Stanford, Penn. State, Los Alamos, Lawrence Livermore, and Phillips Laboratory. We have benefited from stimulating discussions with many colleagues, including Russ Armstrong, Tim Bell, Dennis Boccipio, Gar Bering, Bill Boeck, Bill Feldman, Gerry Fishman, Steve Goodman, Les Hale, Dan Holden, Umran Inan, Phil Krider, Walt Lyons, Gennady Milikh, Dick Orville, Dennis Papadopoulos, Viktor Pasco, Colin Price, Steve Reising, Yuri Taranenko, Skeet Vaughan, Earle Williams, and John Winckler. This work was supported in part by the National Aeronautics and Space Administration under Grant NAG5-5019 and the National Science Foundation under Grant ATM-9217161.

References

Boeck, W. L., O. H. Vaughan, Jr., R. J. Blakeslee, B. Vonnegut, M. Brook, and J. McKune, Observations of lightning in the stratosphere, *J. Geophys. Res.*, *100*, 1465, 1995.

Fishman, G. J., P. H. Bhat, R. Mallozzi, J. M. Horack, T. Koshut, C. Kouveliotou, G. N. Pendleton, C. A. Meegan, R. B. Wilson, W. S. Paciesas, S. J. Goodman, and H. J. Christian, Discovery of intense gamma-ray flashes of atmospheric origin, *Science*, *264*, 1313, 1994.

Franz, R. C., R. J. Nemzek, and J. R. Winckler, Television image of a large upward electrical discharge above a thunderstorm system, *Science*, *249*, 48, 1990.

Holden, D., C. Munson, and J. Devenport, Satellite observations of transionospheric pulse pairs, *Geophys. Res. Lett.*, *22*, 889, 1995.

Lyons, W. A., Characteristics of luminous structures in the stratosphere above thunderstorms as imaged by low-light video, *Geophys. Res. Lett.*, *21*, 875, 1994.

Lyons, W. A., and E. R. Williams, Some characteristics of cloud-to-stratosphere "Lightning" and considerations for its detection, Symposium on the Electrical Circuit, Global Change, and the Meteorological Applications of Lightning Information, American Meteorological Society, Nashville, TN, January 23–28, 1994.

Sentman, D. D., and E. M. Wescott, Video observations of upper atmospheric optical flashes recorded from an aircraft, *Geophys. Res. Lett.*, *20*, 2857, 1993.

Sentman, D. D., and E. M. Wescott, D. L. Osborne, D. L. Hampton, and M. J. Heavner, Preliminary results from the Sprintes94 aircraft campaign: 1. Red sprites, *Geophys. Res. Lett.*, *22*, 1205, 1995.

Sentman, D. D., and E. M. Wescott, Red sprites and blue jets: Thunderstorm-excited optical emissions in the stratosphere, mesosphere, and ionosphere, *Phys. Plasmas, 2,* 2415, 1995.

Wescott, E. M., D. D. Sentman, D. L. Osborne, D. L. Hampton, and M. J. Heavner, Preliminary results from the Sprite94 aircraft campaign: 2. Blue jet, *Geophys. Res. Lett., 22,* 1205, 1995.

Winckler, J. R., Further observations of cloud-ionosphere electrical discharges above thunderstorms, *J. Geophys. Res., 7,* 14,335, 1995.

D. D. Sentman and E. M. Wescott

Geophysical Institute, University of Alaska, Fairbanks, AK 99775.

Magnetic Storms

Bruce T. Tsurutani and Walter D. Gonzalez

Solar Phenomena

One of the oldest mysteries in geomagnetism is the linkage between solar and geomagnetic activity. The 11-year cycles of both the numbers of sunspots and Earth geomagnetic storms were first noted by *Sabine* [1852]. A few years later, speculation on a causal relationship between flares [*Rust*, this vol.] and storms arose when *Carrington* [1859] reported that a large magnetic storm followed the great September 1859 solar flare. However, it was not until this century that a well-accepted statistical survey on large solar flares and geomagnetic storms was performed [*Newton*, 1943], and a significant correlation between flares and geomagnetic storms was noted.

Although the two phenomena, one on the Sun and the other on the Earth, were statistically correlated, the exact physical linkage was still an unknown at this time. Various hypotheses were proposed, but it was not until interplanetary spacecraft measurements were available that a high-speed plasma stream rich in helium was associated with an intense solar flare [*Hirshberg et al.*, 1970]. The velocity of the solar wind increased just prior to and during the helium passage, identifying the solar ejecta for the first time [*Goldstein*, this vol.]. Space plasma measurements and Skylab's coronagraph images of coronal mass ejections (CMEs) from the Sun firmly established the plasma link between the Sun and the Earth. One phenomenon associated with magnetic storms is brilliant "blood" red auroras, as shown in Figure 1.

Types of Solar Wind

Since the early 1960's, plasma and magnetic field instruments onboard interplanetary spacecraft have shown that a continuous flow of plasma

Figure 1. The red aurora created by the emission of 6300 Å oxygen line at very high altitudes (200–600 km) where the collisional de-excitation time scales are larger than the metastable decay time of ~200 s. Courtesy of V. Hessler, Geophysical Institute, University of Alaska, Fairbanks. One thought [*Cornwall et al.*, 1971] is that the electromagnetic ion cyclotron waves generated by the loss cone instability of the ring current protons get damped and accelerate magnetosphere thermal electrons up to energies of ~2–3 eV. These low-energy electrons get stopped high in the atmosphere resulting in the aurora. Another possibility [*Fok et al.*, 1991] is that ring current ions and electrons are slowed down by Coulomb interactions with thermal plasma and are eventually removed from trapped orbits. There is now more evidence supporting this second mechanism.

comes outward from the Sun. At 1 Astronomical Unit (the Earth's distance from the Sun), this "solar wind" has a nominal velocity of ~400 km s^{-1} and a density of ~7 particles cm^{-3}. The plasma consists of primarily hot electrons and protons with a minor fraction (~3–5%) of He^{++} ions. The plasma has an embedded magnetic field of intensity ~5 nT (nanotesla).

Besides the quiescent solar wind discussed above, near solar maximum (maximum number of sunspots), impulsive streams with velocities greater than 600 km s^{-1} and sometimes even greater than 1000 km s^{-1} occur occasionally. Using Newton's 1943 statistics, we know that approximately 90%

of these high-speed streams at solar maximum are associated with ICMEs, the interplanetary component of CMEs. Because the magnetosonic wave speed is only ~60 km s⁻¹, the difference in flow velocity between the faster impulsive stream and the slower stream is greater than the magnetosonic (fast-mode) speed. Thus, a fast forward shock is formed at the leading edge of the high-speed stream.

The shock is the outermost (antisunward) extension of the solar disturbance's propagation into interplanetary space. The region immediately behind (sunward of) the shock is composed of swept-up, compressed, and accelerated plasma and fields from the "slow" stream and is called the "sheath" region. Behind this is the driver-gas (ICME) proper. The driver gas has previously been identified by a variety of signatures: enhanced helium/hydrogen density ratios, low ion temperatures, high-intensity magnetic fields with low variances, and bidirectional streaming of ions and electrons. However, it should be mentioned that no one measurement or combination of measurements has proved to be a perfect means of identification and intense research in this area is still ongoing.

Because of the typically high-intensity magnetic fields and low plasma temperatures, the driver gas is a low-beta plasma, $\beta = 0.03–0.8$. In about 10% of the cases, the magnetic field in these regions has an unusual configuration, with large out-of-the ecliptic components (see Figure 2; *Burlaga et al.* [1990]; *Marubashi* [1986]). This magnetic field structure has been named a magnetic cloud [*Klein and Burlaga,* 1982]. When crossing the cloud, the field rotates from north-to-south or south-to-north with a time scale of a day or longer. This configuration is believed to be force-free, supported only by field-aligned currents flowing inside of it.

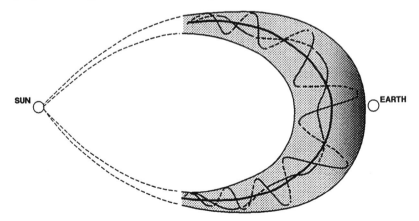

Figure 2. A possible configuration of the magnetic fields within the low beta portion of the driver gas.

Magnetic Reconnection and Magnetic Storms

The high-speed plasma events, which are led by shocks, followed by plasma sheaths and then by the driver gases, do not have direct access to the Earth's dayside atmosphere and ionosphere [*Richmond*, this vol.]. The protective magnetosphere [*Cowley*, this vol.], which is created by the internal magnetic field of the Earth, deflects the interplanetary plasma and fields, so the latter flow around the magnetosphere. The solar wind plasma primarily enters the magnetosphere through magnetic connection between the interplanetary magnetic fields and the Earth's outer fields, as shown in Figure 3. When the interplanetary magnetic field (IMF) has a direction opposite (southward) to the magnetospheric fields (northward), interconnection can take place, and the solar wind convects these fields back into the tail region where they reconnect once more [*Dungey*, 1961]. The magnetic tension on the freshly reconnected tail fields "snaps" the reconnected fields and plasma forward toward the nightside of the Earth. The convection process, through conservation of the first two adiabatic invariants (μ and $\oint \rho_{\parallel} dl$), energizes the plasma. When the magnetic dayside connection is particularly intense, the nightside reconnection is also correspondingly high, and the plasma is driven deep into the nightside inner atmosphere. Because the plasma is anisotropically heated by this process, plasma instabilities (loss-cone instabilities) [*Gary*, this vol.] occur, creating electromagnetic and electrostatic plasma waves, which cyclotron resonate with the energetic particles

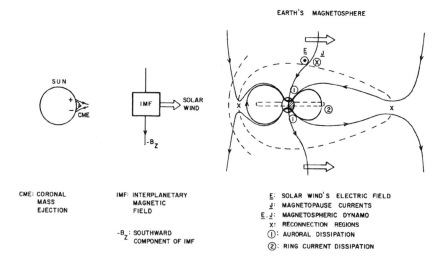

Figure 3. Magnetic reconnection between interplanetary and magnetospheric magnetic fields.

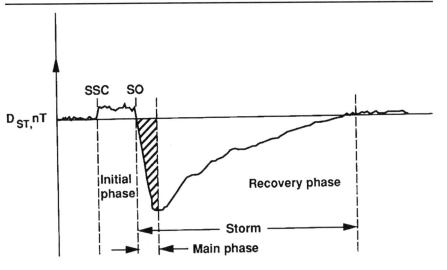

Figure 4. The three phases of a magnetic storm.

[*Kennel* and *Petschek,* 1966]. The wave particle interactions break the particles' first adiabatic invariant, scattering them in pitch angle. Particles that have their mirror points lowered to altitudes at atmosphere/ionosphere heights are lost by collisions with atmospheric/ionospheric particles. In the loss process, atmospheric/ionospheric atoms and molecules are excited, resulting in characteristic auroral emissions [*Akasofu,* this vol.]. The above scenario is the cause of the diffuse aurora, a phenomenon that occurs primarily in the Earth's midnight sector. The spreading of the aurora toward local dawn is caused by electron azimuthal drift [*Cowley,* this vol.].

As the energetic particles are convected deep into the Earth's nightside magnetosphere, they are also subjected to forces due to the magnetic field's curvature and gradient as well as forces due to particle gyration effects. For the same sign charge, these forces act in unison, with the net effect of protons drifting from midnight toward dusk and electrons from midnight toward dawn. This oppositely directed drift comprises a ring of current around the Earth. The current is a diamagnetic one, decreasing the intensity of the Earth's field. An enhanced ring current is the prime indicator of a magnetic storm. The total energy of the particles in the ring current (measured by the intensity of the diamagnetic field perturbation) is a measure of the storm intensity.

We now compare the interplanetary features discussed previously and their relationships to the phases of a magnetic storm, shown in Figure 4, where the ordinate (the field averaged over these ground-based stations near the equator) gives the change in the horizontal component of the

Earth's magnetic field and the abscissa gives time. As indicated in the figure, there are three phases to a geomagnetic storm—the initial phase, where the horizontal component increases to positive values of up to tens of nanoTeslas; a main phase that can have magnitudes of minus hundreds of nanoTeslas; and a recovery phase, where the field gradually returns to the ambient level. The time scales of the three phases are variable. The initial phase can last minutes to many hours, the main phase a half-hour to several hours, and the recovery from tens of hours to a week.

An Interplanetary Example

Solar Maximum

Previously, we showed that a flux-rope configuration could lead to large southward field orientations, magnetic connection at the Earth's magnetosphere (when the IMF is southward), and aurora.

It should be noted that in the sheath and the driver gas, two regions where intense southward interplanetary magnetic fields can occur within high-speed impulsive streams, the field orientation has been found empirically to be northward directed with equal probability as southward orientations. There are also cases where the field lies primarily in the ecliptic plane and cases with large north-south components that vary rapidly in time. The latter cases do not cause storms because of their short reconnection/convection time scales. Therefore, only one in about six cases of impulsive high-speed streams that impinge upon the Earth leads to an intense (D_{ST} < −100 nT) magnetic storm [*Tsurutani et al.*, 1988a]. The above-mentioned driver gas fields apply to storms that occur at or near solar maximum.

Many of the solar wind-magnetic storm relationships discussed above can be illustrated by space plasma data, as shown in Figure 5. From top to bottom, the panels give the solar wind velocity, plasma density, magnetic field magnitude, two components of the magnetic field in Geocentric Solar Ecliptic (GSE) coordinates, the Auroral Electrojet index (AE) and D_{ST}. The auroral electrojet is an ionospheric current that flows at ~100-km altitude and is typically located at auroral latitudes (~63–68° magnetic latitude). The location of this current moves equatorward during magnetic storms. This current becomes particularly intense during active auroral displays and can reach amplitudes of up to 10^7 Amperes. The AE index is a ground-based measurement of the magnetic field associated with this current.

Using the ISEE-3 observations, *Gonzalez and Tsurutani* [1987] found that the ten intense storms (D_{ST} < −100 nT) during August 1978–December 1979 were associated with large-intensity (< −10 nT) and long-duration (> 3 hours) negative B_z events, of the type shown in Figure 5.

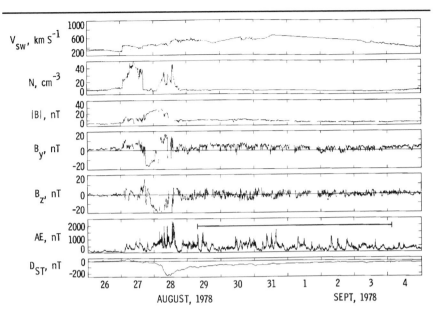

Figure 5. An example of a solar flare-related high-speed interplanetary stream and its geomagnetic effects. Taken from *Tsurutani et al.* [1988b]. An interplanetary shock is noted in the figure at the beginning of August 27 by an abrupt jump in the solar wind velocity, density, and magnetic field magnitude. The increase in ram pressure leads to an increase in D_{ST} to positive values, and is the onset of the storm initial phase. Toward the end of the day, B_z turns negative (southward) and remains in this direction for over 12 hours. D_{ST} decreases in response, signifying the start of the storm main phase, that is, the ring current build-up. As the IMF B_z turns positive (northward), D_{ST} begins to increase, and the onset of the recovery phase begins.

The recovery phase of the storm seen in Figure 5 is exceptionally long. Continuous auroral activity is associated with this interval and is illustrated by the bar in the AE panel. During this time, the interplanetary medium is characterized by rapid fluctuations in the transverse (y and z) components of the magnetic field. The field magnitude is relatively constant. Analyses of the field and plasma data indicate that these fluctuations are Alfvén waves [*Belcher and Davis*, 1971] propagating outward from the Sun. Use of magnetic field measurements on spacecraft closer to the Earth has demonstrated that the AE increases are correlated with southward deviations of the field, the latter associated with the Alfvén waves. Thus, the AE activity is due to magnetic reconnection [*Tsurutani et al.*, 1990; 1995]. However, it is noted that there is very little ring current activity (D_{ST}) during this extended interval.

The lack of ring current activity can be understood by the nature of the southward field components of the Alfvén waves. The fields are less intense than those during the storm main phase (see Figure 5), and their durations are considerably shorter. Thus, the consequential nightside convection will be of lower velocity and will occur sporadically. Plasma will be brought only into the outer regions of the magnetosphere where they feed the high-latitude aurora and not deep into the magnetosphere where the ring current predominantly resides.

Solar Minimum

During the descending phase of the solar cycle, solar coronal holes migrate down to low heliographic latitudes. The continuously emitted high-speed (750-800 km/s) streams emanating from these solar regions interact with lower speed streams in interplanetary space creating regions of compressed magnetic fields. These corotating interaction regions (CIRs) [*Smith and Wolfe*, 1996] can have field intensities of 20-30 nT at 1 AU. They are so named because coronal holes are often long-lasting and the high-speed streams emanating from them and the concomitant CIRs recur every solar rotation.

The CIRs impinging on the earth's magnetosphere have only moderate geoeffectiveness, however. Storms of D_{ST} < -100 nT intensity caused by CIRs are quite rare. The reason is that the IMF B_Z component within CIRs is often highly fluctuating, and the long duration southward IMFs needed for storms are not present.

The Alfvén waves in the high-speed streams following CIRs can lead to exceptionally high auroral zone (AE) activity. It has been shown that this activity during this phase of the solar cycle can be higher than during solar maximum.

Other types of solar wind-magnetospheric interactions, such as a "viscous interaction" between the solar wind and the magnetosphere [*Axford and Hines*, 1961], have been hypothesized. Evidence indicates that the Kelvin Helmholz instability occurs when the IMF is orthogonal (northward) to the tail field direction; however, it was recently shown that only ~0.1% of the solar wind ram energy enters the magnetosphere during these events, compared to 10% during magnetic reconnection intervals (storm events).

Future Space Physics Missions

Where do we go from here? How are we going to fully understand the flow of energy from the Sun to the magnetosphere and the eventual sinks in the ionosphere and magnetotail? The International Solar Terrestrial

Physics (ISTP) mission is devoted to quantitatively solving the energy flow problem discussed in this paper. Scientists from NASA, the European Space Agency, the Russian Space Research Institute, and the Japanese Institute of Space and Astronautical Science will study the energy flow by using data taken from spacecraft placed in interplanetary space (WIND and SOHO), in the magnetosphere (Polar), and in the magnetotail (GEOTAIL).

Acknowledgments. We wish to thank Y. Kamide, L. Lanzerotti, E. Bering, S. Kahler, F. Reese, and R. Thorne for scientific discussions on parts of this paper. The work represented in this paper was performed in part at the Jet Propulsion Laboratory, California Institute of Technology, Pasadena, Calif., under contract with NASA.

References

Axford, W. I., and C. O. Hines, A unified theory of high-latitude geophysical phenomena and geomagnetic storms, *Can. J. Phys., 39*, 1433, 1961.

Belcher, J. W., and L. Davis, Jr., Large amplitude Alfvén waves in the interplanetary medium, 2, *J. Geophys. Res., 76*, 3534, 1971.

Burlaga, L. F., R. P. Lepping and J. Jones, in *Physics of Flux Ropes*, ed. C. T. Russell, E. R. Priest and L. C. Lee, *AGU Monograph 58*, Washington, D.C. 373, 1990.

Carrington, R. C., Description of a singular appearance in the Sun on September 1, 1859, *Mon. Not. Ry. Astron. Soc., 20*, 13, 1860.

Dungey, J. W., Interplanetary magnetic field and the auroral zones, *Phys. Rev. Lett., 6*, 47, 1961.

Fok, M. C., J. U. Kozyra, A. F. Nagy, and T. E. Cravens, Lifetimes of ring current particles due to Coulomb collisions in the plasmasphere, *J. Geophys. Res., 96*, 7861, 1991.

Gonzalez, W. D., J. A. Joselyn, Y. Kamide, H. W. Kroehl, G. Rostoker, B. T. Tsurutani, and V. M. Vasyliunas, What is a geomagnetic storm?, *J. Geophys. Res., 99*, 5771, 1994.

Hirshberg, J., A. Alksne, D. S. Colburn, S. J. Bame, and A. J. Hundhausen, Observation of a solar flare induced shock and helium-enriched driver gas, *J. Geophys. Res., 75*, 1, 1970.

Klein, L. W., and L. F. Burlaga, Interplanetary magnetic clouds at 1 A.U., *J. Geophys. Res., 87*, 613, 1982.

Newton, H. W., Solar flares and magnetic storms, *Mon. Not. R., Astron. Soc., 103*, 244, 1943.

Sabine, E., On periodical laws discoverable in the mean effects of the larger magnetic disturbances, *Philos. Trans. R. Soc. London, 142*, 103, 1852.

Smith, E. J. and J. W. Wolfe, Observation of interaction regions and corotating shocks between one and five AU: Pioneers 10 and 11, *Geophys. Res,. Lett., 3*, 137, 1976.

Tsurutani, B. T., B. E. Goldstein, W. D. Gonzalez and F. Tang, Comment on "A new method of forecasting geomagnetic activity and proton showers" by A. Hewish and P. J. Duffet-Smith, *Planet. Space Sci., 36*, 205, 1988a.

Tsurutani, B. T., W. D. Gonzalez, F. Tang, S. J. Akasofu, and E. J. Smith, Origin of interplanetary southward magnetic fields responsible for major magnetic storms near solar maximum (1978-1979), *J. Geophys. Res.*, *93*, 8519, 1988b.

Tsurutani, B. T., T. Gould, B. E. Goldstein, W. D. Gonzalez, and M. Suguira, Interplanetary Alfvén waves and auroral (substorm) activity: IMP 8, *J. Geophys. Res.*, *45*, 2241, 1990.

Tsurutani, B. T., W. D. Gonzalez, A.L.C. Gonzalez, F. Tang, J. K. Arballo and M. Okada, Interplanetary origin of geomagnetic activity in the declining phase of the solar cycle, *J. Geophys. Res.*, *100*, 21717, 1995.

RECOMMENDED READING: *Magnetic Storms* edited by B. T. Tsurutani, W. D. Gonzalez, Y. Kamide, J. K. Arballo, *Amer. Geophys. Un.* Monograph, 98, 1997.

Bruce T. Tsurutani, Space Physics and Astrophysics Section, Jet Propulsion Laboratory, California Institute of Technology, Pasadena, CA 91109.

Walter D. Gonzalez, Instituto Nacional Pesquisas Espaciais, Sao Jose dos Campos, San Paulo, Brazil.

The Human Impact of Solar Flares and Magnetic Storms

Jo Ann Joselyn

The Sun shines, and Earth and its inhabitants benefit. But the Sun radiates more than light, and these radiations are variable over time scales of seconds to days to years. The consequences for people range from glorious celestial displays—auroras—to subtle but potentially damaging effects on the technological systems that are increasingly important for daily living. For example, electric power transmission systems and communication links have proven vulnerable to solar phenomena. And outside of Earth's protective atmosphere and magnetic shield, there is a small but genuine risk of a solar energetic particle burst that would be lethal to satellite sensors and command and control systems and astronauts.

It has been known since the time of Galileo that the Sun is neither featureless nor steady. Besides ordinary sunlight, there are three classes of solar emanations that can be directly associated with effects at Earth—photon radiation from solar flares [*Rust*, this vol.], solar energetic particles [*Lin*, this vol.], and inhomogeneties in the solar wind [*Goldstein*, this vol.] that drive magnetic storms [*Tsurutani and Gonzalez*, this vol.]. Below, the emanations are summarized and their effects are described.

Solar Flares

Sunspots are dark areas on the solar surface that are transient concentrated magnetic fields [*Foukal*, this vol.]. Groups of sunspots, especially those with complex magnetic field configuration, are preferential sites for "flares." A flare is a spontaneous release of energy (up to 10^{32} ergs) in time spans of seconds to hours. Flares are seen at ground-based observatories as bright areas on the Sun in optical wavelengths and as bursts of noise at radio wavelengths. During solar flares, the flux of photons emitted at X-ray wavelengths (below a nanometer) increases up to several hundred times its usual level. Radiation at wavelengths shorter than 100 nm does not reach the Earth's surface but is absorbed in the upper atmosphere (above approximately 40 km), producing a layer where some of the atoms have been ionized. X-ray wavelengths particularly contribute to ionization at altitudes between 60 and 160 km. Radio waves can be reflected from the ionosphere like light reflects from a mirror. However, the efficiency of reflection depends on the radio wave frequency and the properties of the ionosphere itself, which can change over a time scale of minutes. Extremely high frequencies, such as those used to communicate with satellites, pass right through the ordinary ionosphere. Radio waves from terrestrial sources with frequencies below about 30 MHz are directly affected by the ionosphere. During large flares, which can occur at the rate of several per day during peak activity conditions, all high-frequency (HF) radio waves (3–30 MHz) reaching the daytime ionosphere are absorbed for the duration of the flare. The flare can last minutes to hours. Such radio "blackouts" are significant because HF radio propagation permits communication over long distances, such as short-wave broadcasting by the Voice of America and the BBC as well as by amateur radio enthusiasts. At the same time, the phase of very low wave frequencies (VLF, 3–30 kHz) reflecting from the bottom side of the ionosphere advances as the propagation paths shorten.

Solar Energetic Particles

Most of the time, satellite sensors at geosynchronous altitudes (~36,000 km above Earth's surface) that count protons and helium nuclei with high energies (above approximately 5 MeV/nucleon) measure only "background," a threshold level of instrument noise. However on occasion, significant counts of energetic particles from the Sun are detected, where "significant" is defined as at least 10 particle flux units (pfu) (particles/cm^2-s-sr) with energies above 10 MeV. Proton energies can exceed 100 MeV. The record of the number of significant satellite proton events that have occurred each year since 1976 varies from 1 to 23, and the fluxes range from the event threshold, which is 10, to approximately 40,000 pfu. These events can last from hours to more than a week but typically last 2–3 days. Most proton events—especially those with >10 MeV fluxes exceeding 100 pfu—are associated with flares, but only a small percentage of all flares are associated with proton events.

Energetic particles pose a special hazard at low-Earth orbit and above, where they can penetrate barriers such as spacesuits and aluminum and destroy living cells and solid state electronics. The penetration of high-energy particles into living cells, measured as radiation dose, leads to chromosome damage and, potentially, cancer. Large doses can be fatal immediately. Factors in calculating the hazard to human beings include the composition and thickness of any shielding material and the elemental composition of the impacting particles, which are mainly protons but with some higher-mass particles, as well as the particles' energies and numbers. Solar protons of energies between 10 and 100 MeV are particularly hazardous. In October 1989, the Sun produced enough energetic particles that had there been an astronaut on the Moon, wearing only a spacesuit and caught out in the brunt of the storm, death would have been probable. Astronauts who might have gained safety in a shelter beneath moon soil would have absorbed only slight amounts of radiation—on the order of the lifetime dose of an average, non-spacefaring citizen. The October 1989 event was so extraordinary that it produced elevated dosimeter readings—but less than the annual dose limits set for the general public—onboard supersonic transports flying at high altitudes over the polar caps. At sea level, there was no noticeable radiation increase because the Earth's atmosphere has the absorption equivalent of 10 m of water. However, over their lifetimes, satellites at high altitudes or in low-altitude polar orbits accumulate doses many times that of the lethal human limit with results ranging from damaged surface materials to logic circuit upsets in computer memories and control mechanism. Energetic particle events like those of October 1989 are relatively rare, but they are, as yet, unpredictable. The most recent one of a comparable size occurred in March 1991. The only strategies for mitigation are shielding, careful selection of designs and materials, and the use of redundancy and self-checking in logic systems.

Energetic solar particles also influence terrestrial radio waves propagating through polar regions in a separate process than the one caused by solar flare X-ray radiation, which affects only the sunlit side of the Earth. Although energetic particles are shielded from lower latitudes by the Earth's magnetic field, they gain access to the ionosphere over the polar caps, where the magnetic field shielding is less effective. Polar cap absorption (PCA) events are troublesome to radio navigation techniques making use of the nearly constant height of reflection of very low frequency waves to find the propagation time, and hence the distance, to the beacon. During PCA events the height of reflection lowers. Positioning errors on the order of kilometers are possible on transpolar paths if a PCA event is unrecognized.

Magnetic Storms

Although flare radiation and solar energetic particles have important and noticeable effects, the most pervasive human effects can be attributed to magnetic storms, which are the response of the Earth's magnetic field to specific inhomogeneities in the solar wind. Stated simply, the solar wind is the expansion and escape of the outer solar atmosphere into interplanetary space. Inhomogeneities arise because the outer solar atmosphere—the corona—is structured by the strong solar magnetic fields. Where the magnetic field is relatively weak or aligned with the gravity field, the atmosphere can readily escape, leading to high-speed solar wind streams. In other places, i.e., above sunspot groups, closed magnetic fields impede or confine solar wind flow. However, the confined atmosphere can be released in bubbles or tongues of plasma and magnetic fields called coronal mass ejections (CMEs). CMEs add to the complexity of the ambient solar wind. They are associated with some flares, but more often they occur independently from flares. These structures in the solar wind have been observed and identified by ground-based optical and radio measurements and in situ spacecraft measurements.

As observed near Earth, solar wind speeds are typically about 400 km/s, but speeds exceeding 1000 km/s have been measured. Proton and electron number densities are typically near 5 cm^{-3} but occasionally exceed 100 cm^{-3}. The solar wind flows around obstacles such as the planets, but those planets with intrinsic magnetic fields respond to the solar wind in a specific way. In effect, Earth's magnetic field activity senses and reacts to the solar wind—its speed, density, and magnetic field. Because the solar wind is variable over time scales as short as seconds, the interface that separates interplanetary space from the magnetosphere [*Cowley*, this vol.] is remarkably dynamic. Normally, this interface—the magnetopause—lies at a distance equivalent to about 10 Earth radii in the direction of the Sun. However, during episodes of elevated solar wind density or velocity, the magnetopause can be pushed inward to within geosynchronous altitudes (6.6 Earth radii). As the magnetosphere extracts energy from the solar wind, internal processes produce geomagnetic storms, increase the probability of auroras at low latitudes [*Akasofu*, this vol.], and change the properties of the ionosphere and upper atmosphere [*Richmond*, this vol.].

Geomagnetic storms are extraordinary variation, albeit only a small percentage, in the surface magnetic field. Consider the compass—a rudimentary instrument that reveals the direction of the Earth's magnetic field. Workers have learned to make use of sophisticated compasses called magnetometers to assist with navigation and geophysical exploration. To those workers, even moderate fluctuations in Earth's relatively steady field are a concern.

Further, rapidly fluctuating fields induce currents in long "wires" (i.e., power lines, pipelines, cables, and even train tracks) that have led to equipment failures in the past. A recent case is the loss of the power grid in

Quebec, Canada, on March 13, 1989; 6 million people were without commercial electric power for 9 hours. The geomagnetic storm that bears responsibility for that outage produced total deviations in compass heading of several degrees even at middle latitudes. Was the March 1989 storm singular in its intensity? It was the largest in recent memory, but in the years since 1868, the first year of the longest series of geomagnetic index records, comparable or larger storms were observed on September 25, 1909; September 18, 1941; and November 13, 1960.

An associated consequence of Earth's response to blasts of solar wind is the energization of a population of electrons and ions resident in the magnetosphere. These trapped particles, guided by the roughly dipolar geomagnetic field, usually enter the upper atmosphere near the polar regions. They strike the molecules and atoms of the thin, high atmosphere, exciting some of them to glow. These are auroras, dynamic and delicate displays of colored light seen in the night sky. The incoming particles deposit their energy in the neutral atmosphere, heating it. The heated "air" rises, and the density at the orbit of satellites up to about 1000 km increases significantly. As a result of the added frictional drag, satellites lose energy and their orbits change. All low altitude satellites are slowly falling back to Earth owing to atmospheric drag; this process is accelerated during geomagnetic storms. For example, the NASA Long Duration Exposure Facility satellite, which was recovered from an altitude of 340 km in January 1990 lost 500 m of altitude in 1 day as a result of the March 1989 storm. Normal daily loss rates were about 200 m/day at that time. Individual satellites respond differently to increased density, so agencies that monitor the positions and identify the approximately 6000 objects in low-Earth orbit, satellites and debris larger than about 10 cm in diameter, require additional resources during magnetic storms. Lower altitude navigation satellites can be affected to the point that they are useless until their new orbits stabilize.

During magnetic storms, some of the energized magnetospheric particles are trapped above the tangible atmosphere; they circulate around Earth and form a ring of current [*Van Allen*, this vol.] that can be sensed on the ground by its associated magnetic field. When these energized particles impact satellites—especially at geostationary orbits—portions of the satellite surface can charge up. Differential charge exceeding 10,000 V has been measured, and arcing can occur. The effects of arcing on the satellite are not predictable and can be damaging. Over time, the physical properties of the surface materials can be altered.

Another result of the energy deposited in the upper atmosphere during geomagnetic storms is ionospheric storms. Like flares, ionospheric storms affect radio communication at all latitudes, but these storms last for hours to days and disturb frequencies from 3 kHz to 30 GHz. Patches of transient ionization occur as a strong function of latitude and time of day. Some frequencies are absorbed and others are reflected, leading to anomalous propagation paths and rapidly fluctuating signals. Long-range radars experience unusual signal retardation and refraction, causing distance and pointing errors. Even satellite communication systems operating through the disturbed ionosphere may experience phase and amplitude scintillations.

Finally, there is a growing body of evidence that changes in the geomagnetic field affect biological systems. In particular, homing pigeons and other migratory creatures appear to use the magnetic field as at least a backup navigational aid. Other studies indicate that physically stressed human biological systems may respond to the minute but measurable fluctuations of the geomagnetic field. Interest and concern in this subject contributed to the decision by the Union of Radio Science International to create a new commission entitled "Electromagnetics in Biology and Medicine."

It has been realized and appreciated only in the last few decades that solar flares and magnetic storms affect people. The list of consequences is growing in proportion to our dependence on technological systems. The subtleties of the interactions between the Sun and Earth and between solar particles and delicate instruments have become factors affecting our well-being.

Recommended Reading

Davies, K., *Ionospheric Radio*, London: Peter Peregrinus, 1990.

Eather, R. H., *Magestic Lights*, Washington, D.C.: AGU, 1980.

Garrett, H. B., and C. P. Pike, eds., *Space Systems and Their Interactions with Earth's Space Environment*, New York: American Institute of Aeronautics and Astronautics, 1980.

Gauthreax, Jr., S. A., *Animal Migration, Orientation, and Navigation*, Chapter 5, New York: Academic Press, 1980.

Harding, R., *Survival in Space*, New York: Routledge, 1989.

Johnson, N. L., and D. S. McKnight, *Artificial Space Debris*, Malabar, Florida: Orbit Book Co., 2987.

Lanzerotti, L. J., ed., Impacts of ionospheric/magnetospheric process on terrestrial science and technology, in *Solar System Plasma Physics*, Vol. III, Edited by L. J. Lanzerotti, C. F. Kennel, and E. N. Parker, New York: North Holland Publishing Co., 1979.

Parkinson, W. D., *Introduction to Geomagnetism*, New York: Scottish Academic Press Ltd., Elsevier Science Publishing Co., 1983.

Jo Ann Joselyn
Space Environment Center, NOAA, 325 Broadway, Boulder, CO 80303

The Solar Wind

B. E. Goldstein

The first evidence of the solar wind was provided through observations of comet tail deflections by L. Biermann in 1951. A cometary ion tail is oriented along the difference between the cometary and solar wind velocities, whereas the dust tail is in the antisunward direction; the ion tail directions demonstrated the existence of an outflow of ionized gas from the Sun (the solar wind) and allowed estimates of solar wind speed. Spacecraft observations have now established that at 1 AU the solar wind has a typical ion number density of about 7 cm^{-3} and is composed by number of about 95% protons and 5% Helium, with other minor ions also present. The solar wind as observed at 1 AU in the ecliptic has speeds typically in the range 300–700 km/s. At such speeds ions travel from the Sun to 1 AU in from 2.5 to 6 days. The impact of the solar wind on planets with magnetic fields (Earth, Jupiter, Saturn, Uranus, Neptune) causes phenomena such as magnetospheres [*Cowley*, this vol.], aurorae [*Akasofu*, this vol.], and geomagnetic storms [*Tsurutani and Gonzalez*, this vol.], whereas at objects lacking magnetospheres (Mars, Venus, comets), atmospheric neutrals undergo charge exchange and are picked up by the solar wind flow. The solar wind also shields the Earth from low energy cosmic rays [*Jokipii*, this vol.], and is responsible for the existence of the anomalous component of the cosmic rays [*Mewaldt et al.*, this vol.] a low energy component that is created locally rather than in the galaxy. Presented here is a brief introduction to the solar wind and a description of some current topics of research; for the current status of solar wind studies see, e.g., Solar Wind 7 [Proc. 3rd COSPAR Colloquium, E. Marsch and R. Schwenn, Ed., Pergamon, 1992].

Solar wind properties vary a great deal due to the changing magnetic structure on the Sun [*Rust*, this vol.]. Large scale streams comprised of high speed, high temperature plasma are observed for periods of several days in the solar wind. These solar wind streams are caused by magnetic structures in the corona that rotate with the Sun. Changes in the solar magnetic field can cause sudden alterations of the balance of forces within the corona, leading to transient events in the solar wind. Large masses of plasma, extending over as much as 40° of the solar disc, are seen to be accelerated outwards from the corona over periods of hours; these events are known as coronal mass ejections. Coronal mass ejections also at times cause shocks which can produce fluxes of energetic particles [*Lin*, this vol.] and cause geomagnetic storms [*Tsurutani and Gonzalez*, this vol.]. On longer time scales, the solar wind is known to vary with the 11-year solar cycle [*Hathaway*, this vol.] and cause geomagnetic storms [*Tsurutani and Goldstein*, this vol.]. The impact of cosmic rays on the Earth's atmosphere produces C^{14}. C^{14} data and historical auroral observations have established that the solar wind varies on time scales much longer than the solar cycle. For example, during the period from 1630 to 1710 very few sunspots were observed (the Maunder minimum); increased C^{14} production occurred at that time.

Origin of the Solar Wind

In simplest terms, the solar wind exists because the solar corona is hot and the pressure in the local interstellar medium is far less than that in the corona. The mechanisms that heat the corona and accelerate the solar wind are the subject of considerable debate, even after 50 years of awareness that the corona is far hotter than the Sun's surface. The corona, a hot ionized gas with a temperature of over 10^6 K, is just above the photosphere (6000 K); radiative and conductive cooling would quickly eliminate the corona were it not for strong coronal heating. Coronal heating on magnetic field lines that are closed (magnetic loops with both feet embedded in the Sun, see Figure 1), is better understood than heating on open magnetic field lines that extend into interplanetary space. Closed coronal magnetic field lines are continuously twisted by the random walking of their photospheric "footprints" due to convection in the outer layers of the Sun; the energy stored in the field dissipates and thereby heats the coronal plasma. However, on open field lines, since the far end of the field line is not tied to the Sun, such twists in the magnetic field can propagate as Alfvén waves (magnetohydrodynamic shear waves) into the solar wind. Dissipation of these Alfvén waves by various nonlinear or resonant processes has been proposed to heat and accelerate the solar wind, but coronal observations indicate that the bulk of the heating occurs within two solar radii of the solar surface while the Alfvén waves are predicted to dissipate over longer length scales. Possibilities that have been suggested for heating in the lower corona include dissipation of

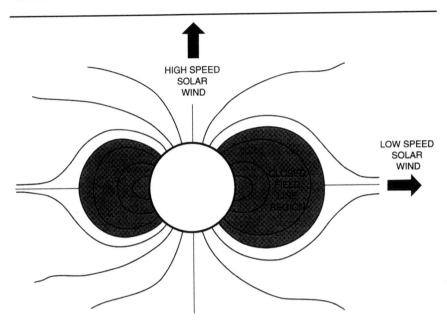

Figure 1. The corona is trapped on closed magnetic field lines (high density) but escapes as the solar wind on open field lines (low density coronal hole regions). The open field lines are typically found at higher latitudes.

energy by reconnection of twisted magnetic fields wherein the magnetic energy could be released in the form of microflares, injection of ion beams, and waves modes that are rapidly damped (magnetoacoustic waves).

In view of the complexity of the processes acting in the region above the solar surface, more direct observational evidence is needed to resolve this fundamental question of heating not only the solar atmosphere, but of all related stellar atmospheres. Two space missions offer possibilities for major advances. Today's telescopic observations can not resolve the small scale activity in the chromosphere and corona that is suspected to provide a major portion of low coronal heating. Multiwavelength optical observations with 0.1-arcsec resolution are required and must be obtained above the atmosphere because daytime atmospheric turbulence limits resolution. A high resolution solar observatory in Earth orbit designed to measure small scale activity would provide needed information for coronal and solar wind research. A proposed Solar Probe mission with perihelion three solar radii above the solar surface would obtain direct in situ measurements of particle distributions in the corona, and much more sensitive measurements of small scale x-ray flare activity than can be obtained from Earth orbit. Such data is the only way to determine whether injected ion beams that could cause coronal heating are present, and determine whether the small scale activity that results in heating is impulsive and related to magnetic reconnection (which would produce x-ray emission), or instead results from other dissipative processes.

Inferences about the heating and acceleration of the solar wind presently rely upon in situ observations of plasmas in interplanetary space. Solar wind observations have been obtained from 0.3 AU to beyond 50 AU, and a wide variety of processes operate within the solar wind and affect its evolutions.

Attempts to understand solar wind physics utilize a wide variety of information including observations of bulk properties of the solar wind, fluctuations, the nonthermal velocity distributions of the ions and electrons, the composition of the solar wind, magnetic connection to the Sun determined from electron and energetic particle data, and plasma wave observations and telescopic observations over a range of wave lengths from radio through x-ray. The physical processes being studied include plasma wave instabilities, thermal conduction in systems for which the mean free path is comparable to the scale size of the system, collisionless shocks and associated acceleration of energetic particles, and turbulence. The solar wind provides an excellent laboratory for magnetohydrodynamic turbulence over a wide range of spatial scales with the additional challenge of understanding the role of collisionless damping of some of the wave modes.

Double proton beams are observed frequently at 1 AU in high speed streams originating in coronal holes which are low density regions of the corona; the density is low because the regions are magnetically open and the coronal plasma escapes as the solar wind. The multiple beams may have a coronal origin in a transient acceleration process, or may instead result from the evolution with increasing distance from the Sun of a high speed beam for which collisions with other protons are unimportant. Interestingly, double proton streams are not observed in other regions such as coronal mass ejections. Additionally, alpha particles in the solar wind are observed at times to be traveling faster than protons; this velocity difference is probably due to wave acceleration, but is not well understood at present. In situ observations closer to the Sun should reveal the cause of these nonthermal distributions.

The solar wind is composed differently than the Sun. Ions are created in the chromosphere, and then lifted into the corona by magnetic and electric fields. Higher relative abundances are observed for atoms that are easy to ionize (i.e., Fe, Mg) than for atoms with higher ionization potentials (i.e., He, Ne), but the details of how the ion selection works are not known. Previous to the Ulysses mission, solar wind experiments typically measured only energy per unit charge, so in many cases different species and ionization states could not be distinguished. The Ulysses ion mass spectrometer has found striking correlations between the abundances of certain species and ionization states that are created in the corona, other species and ionization states that are created in the transition region, and the solar wind speed, providing new constraints on models of the heating and mass separation processes that produce the solar wind.

The High Latitude Solar Wind

The solar wind flows from open magnetic field lines, which usually connect to the Sun at high latitudes; this nonspherical expansion causes latitudinal gradients including a higher velocity over the solar poles. Ulysses plasma observations have shown that the solar wind speed over the poles during solar minimum is typically about 750 km/s with small variability, almost twice as large as the speed of the in-ecliptic plasma that comes from the low-latitude coronal streamer belt. Additionally, because the foot of a magnetic field line rotates with the solar surface, outward solar wind flow produces a spiral field at low latitudes and an approximately radial field at high latitudes (Figure 2). This magnetic field geometry has important consequences for both the solar wind plasma and cosmic rays. In a collisionless plasma the temperature parallel to the magnetic field (parallel temperature is based upon the velocity components of the particles in the direction parallel to the magnetic field) is generally not equal to the perpendicular temperature. For the solar wind a more radial magnetic field (high latitude case) means a larger temperature in the direction parallel to the magnetic field than for the case of a spiral field (low latitude case); the effect upon the perpendicular temperature is the opposite. The altered anisotropy in turn affects the thermal conductivity of the plasma and the plasma instabilities that may be present. Cosmic rays, like all other charged particles, must to a first approximation travel along magnetic field lines. Energetic particles entering the solar system over the poles encounter radially oriented field lines, whereas energetic particles entering in the equatorial region must travel a much longer distance along the spiral field to move inward an equal radial distance. Just as a gas is refrigerated by expansion, cosmic rays lose energy by scattering from magnetic irregularities in an expanding solar wind. Thus, more rapid access over the poles had been expected with less energy loss and consequently larger cosmic ray fluxes. However, Ulysses did not find large increases in the cosmic ray flux at high latitudes, and other Ulysses observations provided the explanation: large amplitude magnetohydrodynamic waves were observed at high latitudes that scatter the cosmic rays and prevent rapid access over the poles.

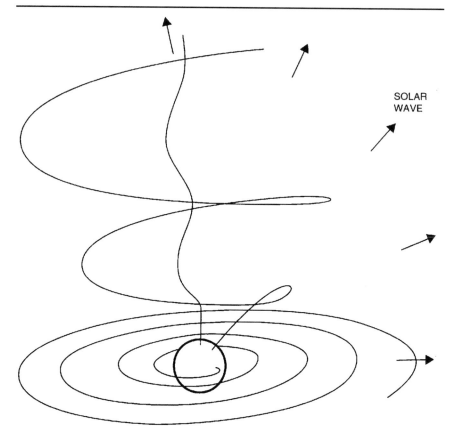

Figure 2. The rotation of the Sun causes the outward convecting magnetic field to have a spiral geometry in the solar equatorial plane; at high latitudes the field is approximately radial.

Outer Heliosphere and Termination Shock

Interest in the outer heliosphere has steadily been increasing as the Voyager and Pioneer spacecraft explore the region beyond 50 AU; evidence has emerged that the heliospheric termination shock [*Axford and Suess*, this vol.] is being approached. The heliospheric termination shock is the boundary at which the solar wind is decelerated to subsonic speeds so that it can turn and flow downstream in the local interstellar medium (LISM). The LISM plasma and magnetic field directly exert pressure on the solar wind, and the interstellar neutral population is important in two ways. First, for both the shocked solar wind and interstellar ion populations, charge exchange with the neutrals can be important. The effect of charge exchange is to cool shocked plasma and couple the momentum of the neutrals to the flow. Closer to the Sun where the solar wind has not yet reached the shock, another major consequence results from charge exchange with the interstellar neutrals; the newly created ions (pickup ions) are accelerated by the solar wind electric field into cycloidal orbits with thermal speed equal to the solar wind outflow speed. The pickup ions thus have much more thermal energy per particle than the colder solar wind plasma. These pickup ions can then be accelerated to much higher energies by processes associated with the termination shock and may alter the structure of the shock.

How far from the Sun is the heliospheric termination shock? The distance could be calculated if the properties of the LISM were known. Recent in-situ detection of interstellar neutral helium atoms by the Ulysses neutral gas experiment, detection of interstellar hydrogen pickup ions by the Ulysses ion mass spectrometer, and optical measurements have established the density, velocity and temperature of neutral gas in the LISM. Unfortunately, the density of ions in the LISM is highly uncertain, and, even worse, there is almost no knowledge of the LISM magnetic field. So, observation of the heliospheric termination shock will provide information about the properties of the LISM, and it possibly will provide our first example of an energetic particle mediated shock. Interested parties hope that the shock will be traversed before they or the Voyager spacecraft die or lose power (the Pioneer 10 and 11 spacecraft will not detect this event because of trajectory and power). Recent evidence suggests that the termination shock may be within range. The Voyager spacecraft have recently detected plasma waves that are perhaps being generated at the termination shock. Additionally, the Voyager and Pioneer spacecraft have both seen an increase of roughly 10% (energy dependent) per AU in the anomalous cosmic ray energy population. If the trend continues, by 80 to 100 AU the energy density of the anomalous cosmic ray population would equal the dynamic pressure of the solar wind. This suggests that energetic particles may play an important role in decelerating the solar wind, with the solar wind being decelerated over a broad region by a cosmic ray pressure gradient, with an embedded conventional shock probably present. This scenario has never been observed in the laboratory or in space, yet is a reasonable hypothesis for the solar wind termination shock, and shocks at supernovas responsible for cosmic rays' acceleration.

Reference

Marsch, E., and R. Schwenn (Eds.), Solar Wind 7-Proceedings of the 3rd COSPAR Colloquium, Pergamon Press, New York, 1992.

B. E. Goldstein
Jet Propulsion Laboratory, California Institute of Technology, Pasadena, CA 91109

Solar Flares

David Rust

> "From veils of haze now comes the gleam,
> Here to a tender scarf it tapers,
> Here gushes forth a vivid stream;
> Then threads of light in a network surging
> Their silver veins through valleys run,
> Till, gathered by the hills converging,
> The sundered filaments are one."
>
> *Goethe: Faust*

The Sun is constantly changing. Not an hour goes by without a rise or fall in solar x-radiation or radio emission. Not a day goes by without a solar flare. Our active star, this inconsistent Sun, this gaseous cloud that blows in all directions, warms the air we breathe and nourishes the food we eat. From Earth, it seems the very model of stability, but in space it often creates havoc.

Over the past century, solar physicists have learned how to detect even the weakest of solar outbursts or flares. We know that flares must surely trace their origins to the magnetic strands stretched and tangled by the roiling plasma of the solar interior. Although a century of astrophysical research has produced widely accepted, fundamental understanding about the Sun [*Foukal*, 1990], we have yet to predict successfully the emergence of any magnetic fields from inside the Sun or the ignition of any flare.

As in any physical experiment, the ability to predict events not only validates the scientific ideas, it also has practical value. In astrophysics, a demonstrated understanding of sunspots, flares, and ejections of plasma would allow us to approach many other mysteries, such as stellar X-ray bursters, with tested theories.

Accurate predictions of solar activity will allow us to use space with less risk and cost. Solar bursts can cripple satellites or shorten their lifetimes and can disrupt communications and electric power distribution [*Joselyn*, this vol.]. They bombard the upper atmosphere with high-energy particles, potentially threatening polar air travelers with unacceptable levels of radiation. In deep space, beyond Earth's magnetosphere, flare radiation can be very dangerous. Unless astronauts can be warned of impending hazards, manned voyages to the planets may be unthinkable. Given adequate warning, astronauts can find temporary shelter behind thick aluminum plates, but they cannot spend the whole voyage in such armor. They will need reliable solar flare forecasts.

Preflare Activity

Spots on the Sun's bright surface, the photosphere, signal the emergence of strong magnetic fields. We think that the collisions, twisting, and eruptions of these fields produce the ten or more flares each day at the maximum of the 11-year sunspot cycle. What is fundamentally not understood about solar flares is exactly what sets them off. That their energy comes from magnetic tension is well established, but what releases the tension? Is it the accidental collision of fields, or is it a sudden tug or push by the plasma below the surface? Or does the eruption of a flare mean that some systematic process makes magnetic fields unstable?

Perhaps the most exciting recent development in our understanding of preflare activity is the discovery of a steady coronal bulging that begins days before a major eruption. The arcades of plasma-filled loops and streamers in the solar corona above the site of the coming flare expand as though the magnetic fields were extending their region of influence. After several days of growth, the whole arcade seems to erupt. The speed sometimes exceeds 1000 km/s. A massive bright filament (Figure 1) frequently accompanies the eruption. Intense electromagnetic emission and atomic particle acceleration follow.

Not all flares start with a coronal mass ejection. Some flares seem to be confined in closed loops that do not erupt into the outer corona. Confined flares generally brighten and fade in less than an hour—much more rapidly than eruptive flares, which sometimes sustain an X-ray-emitting, 10,000,000 K plasma in the corona for a day or longer.

Most flares occur in twisted or stressed magnetic fields, which can be mapped by modern solar magnetographs. These instruments analyze the polarization of the light in high-resolution pictures of sunspot groups. From the polarization, we infer the direction and strength of the magnetic fields. The fields in flare-producing active regions may appear stable for

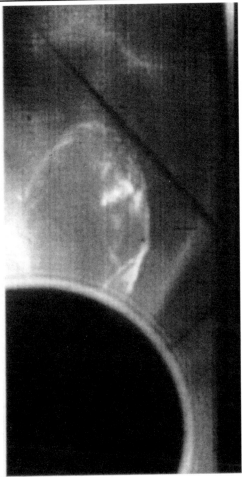

Figure 1. Image of a coronal mass ejection and filament eruption on January 30, 1989, obtained with the white-light coronagraph on the Solar Maximum Mission (courtesy A. Hundhausen).

days, but in the hours before flare onset, the fields in some parts of the region change, often rather dramatically. Classification of these changes is only beginning, because most magnetographs have only come on-line in the past few years.

The bulging magnetic loops in the corona may result from twists propagating upward from the underlying photospheric fields. Because most energy resides in the turbulent photosphere, we expect the photospheric fields to twist the coronal fields, where flares start, but observations have not yet detailed the process. Simultaneous observations of the magnetic fields and the loops are needed. The marvelous images of coronal loops now being obtained with the X-ray telescope on the Japanese satellite Yohkoh are likely to answer many questions.

Flare Phenomena at the Sun

In photographs of the chromosphere, which lies between the diaphanous corona and the dense photosphere, a flare is a few glistening rivulets flowing around a whirl of sunspots. In an X-ray picture of the corona, it is a blinding loopy cloud. By the strictest definition, a flare is the burst of light, X rays, gamma rays, and radio emission from the ribbons and loops, but we now understand that most of the energy in a flare is not in the radiation. Ninety percent of the energy is in the motions of the expanding, twisting magnetic loops that arch out into the solar atmosphere just before the first burst of electromagnetic radiation. But the preflare loop expansion is hard to detect, and the flare radiation is still an important tracer of the physical processes.

Visible flare emission often starts in a few intensely bright knots near sunspots that expand to form two or more ribbons engulfing the spots. The largest flare ribbons cover an area on the solar surface of over a billion square kilometers. Usually, an arcade of brilliant X-ray-emitting loops flare up in the corona just above the ribbons. The ribbons and loops fade in about an hour.

Some eruptive flares make a permanent change in the corona. One of the large streamers, so prominent, for example, in eclipse pictures, may disappear. Also, the chromosphere may eject a dark filament (Figure 2a). An examination of "flare patrol" films will frequently show that a filament thickened half an hour before the flare and twisted and erupted into space, as seen in Figure 1. The fact that the filament eruption and coronal mass ejection start before the radiation suggests strongly that flares result from magnetohydrodynamic instabilities.

The instability probably opens magnetic fields that atomic particles can easily follow into interplanetary space. Their acceleration is still not understood but may take place in shocks generated where the opened fields begin to reconnect and collapse back to the Sun [*Lin*, this vol.]. The signature of reconnecting fields is intense heating, which can drive the coronal temperature to 100,000,000 K and maintain it there for an hour or longer. Thermal conduction fronts or beams of electrons carry energy to the chromosphere to produce the optical and ultraviolet emissions by which flares are best known. The latest results from Yohkoh have not yet resolved a long-standing controversy about whether conduction or beams predominate.

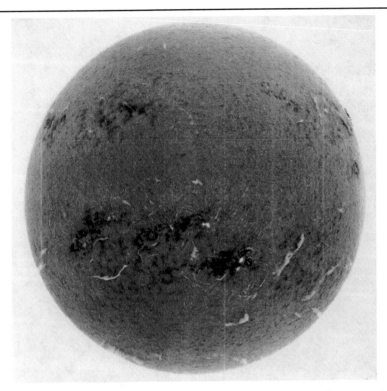

Figure 2a. The Northern Hemisphere of the Sun on July 29, 1967, when it was heavily populated with dark filaments.

Atomic Particle Acceleration

Most flares are atomic accelerators, particularly of protons. A plot (Figure 3) of the X-ray emission from a series of big flares shows the Sun becoming 100–1000 times brighter before each corresponding proton stream reaches Earth. Flare protons can arrive in 20 minutes or in several hours. Often, the flux of 10-MeV protons stays at a dangerously high level for hours.

By monitoring flares with a network of telescopes and satellite-borne detectors, forecasters at the National Oceanic and Atmospheric Administration's Space Environment Forecast Center try to warn of impending proton storms. Their forecasts are based primarily on statistical records from thousands of past flares. But statistics is not destiny, so even a large flare may produce either no proton storm or one aimed at another part of the solar system. The protons are invisible, and their route through space can only be inferred from the way they strike satellite proton detectors.

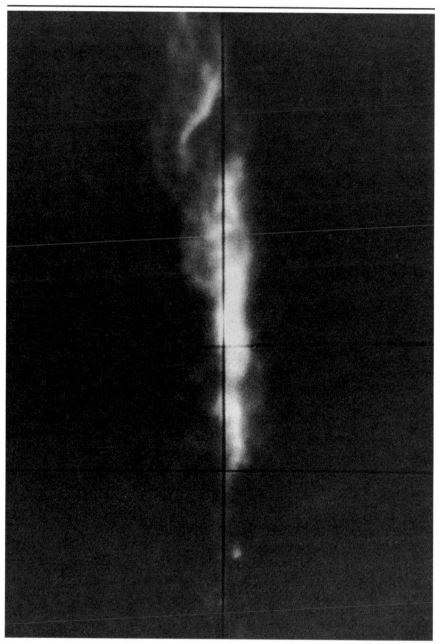

Figure 2b. When a filament erupts, it is bright against the dark sky and often has a helical shape, as shown here. After about half an hour, it may appear as a bright helix in a coronagraph image, as in Figure 1.

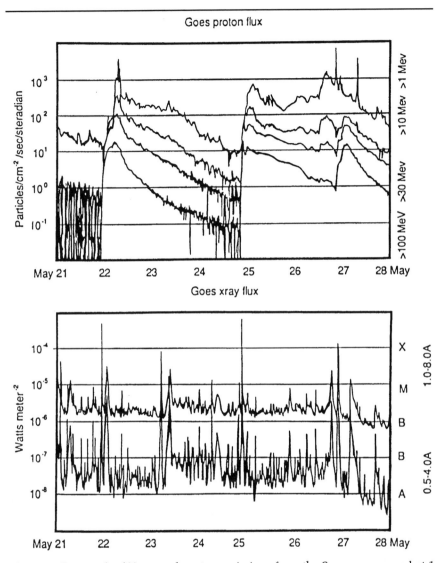

Figure 3. One week of X-ray and proton emissions from the Sun, as measured at 1 AU with the NOAA GOES satellite.

Are Flares Necessary?

The events that lead to an instability may occur on a small scale (magnetic flux cancellation), an intermediate scale (flux emergence), or a large scale (twisting). No one knows whether any of these terms describe a process that is both necessary and sufficient to destabilize filaments and coronal arcades, but recent evidence from measurements of fields in filaments and in interplanetary space suggests that flares may be necessary from a global point of view to relieve the Sun of twisted magnetic fields.

Satellite-borne magnetometers show that the solar wind is disturbed occasionally by magnetic clouds, which are regions ~0.25 AU across (Astronomical Unit = distance from the Sun to Earth) and in which the magnetic-field strength is higher than average. The measured field direction at 1 AU rotates when a cloud passes, suggesting that the field is twisted, as it is in filaments (Figure 2b).

To take advantage of the potential for in-situ magnetic-field measurements in filaments (if that is what the clouds are), I searched for reports of timely filament eruptions within 45° of the solar disk center. Each candidate eruption occurred about 4 days before the arrival at Earth of a well-defined magnetic cloud. The precise interval of search for each event depended on cloud velocity and the solar wind velocity at 1 AU, since the transit time is approximately 1 AU divided by the cloud speed. I found six candidate filament eruptions without flares and ten flares having some evidence of an associated filament eruption.

The fields in all the clouds, as measured by *Lepping et al.* [1990], were twisted into helices. In thirteen of sixteen clouds the fields had a sense of twist (chirality) that agreed with the usual sense for the fields in the corresponding solar events, that is, left-handed in the northern hemisphere, right-handed in the southern hemisphere. Eight clouds were left-handed, eight were right-handed. Seven filaments came from the South, nine from the North. Thus, the probability that thirteen or more coincidences would be found if the twist direction of filaments were random is 0.01.

Vrsnak et al. [1991] showed, from a separate study of fifteen stable and thirteen eruptive filaments with helical structures, that whenever a filament showed a net twist, that is, apparent end-to-end rotation of the band-shaped filament, of ~2.5 π radians or more, it arched outward and erupted. We might consider 2.5 π a quantum of twist, since many theoretical studies and now an observational study have shown that filaments erupt when the fields in them exceed that degree of twist.

It has long been thought that turbulence or the shearing effect of differential plasma motions in the photosphere could somehow produce enough twist to destabilize filaments. However, motions that produced a twist of 2.5 π rarely, if ever, happen on the surface. The ultimate twisting forces are probably beneath the photosphere.

From maps of the rotating layers beneath the photosphere, we can guess at the amount of twist for submerged flux ropes in the northern and southern hemispheres. All the layers rotate faster at the equator, and by doing so, they stretch submerged magnetic flux ropes into toroids. The layers just beneath the surface also rotate faster. We can speculate that the aerodynamic drag of this motion twists the flux ropes into right-handed helices in the north and left-handed ones in the south.

The north-south pattern of twists on the Sun is always the same, regardless of magnetic-field direction and regardless of solar-cycle number. This pattern has been found for sunspot whirls, twisted filaments, and now in the study of magnetic clouds described above.

Magnetic helicity is extremely difficult for solar plasma to shed in place. It is not surprising, therefore, that the helical shape seen in erupting filaments is still detectable 4 days later when the plasma drifts past the Earth. Solar magnetic fields could no more spontaneously relieve their twist than could steel cables. All the evidence indicates that the subsurface motions are continually twisting the fields. In each hemisphere, the twist is always in the same direction. So why doesn't the Sun tie itself in knots?

Suppose the flux of ~10^{21} Maxwells (Mx = 10^{-8} weber) at the poles at the beginning of a solar cycle is divided into flux ropes, each with ~3×10^{17} Mx, which is the observed flux in the ropes. Then, the number of ropes crossing the equator, where the jet stream is, would equal ~3000. In the submerged equatorial jet stream, each layer will advance by ~12 km/day in the direction of rotation beyond the layer 100 km next to it. Suppose each flux rope is efficiently twisted by the aerodynamic drag in this layered motion. Then each acquires 2.5 π of twist every 63 days. Since we suppose there are 3000 flux ropes, we expect about fifty eruptions per day, each shedding a quantum of twist. Now we may compare the helicity spawning rate by field-wrapping in the subsurface jet stream with the shedding rate in eruptions. In fact, there are about ten eruptive events per day on the visible hemisphere, so the speculation that the Sun sheds helicity by eruptive flaring is at least plausible, considering that flux ropes are probably not twisted as efficiently as in this example.

Our line of reasoning suggests that solar flares are not the result of accidental collisions or instabilities, but rather that they are necessary to purge magnetic helicity from each hemisphere. The magnetic cloud study suggests that the twist generated by solar rotation is thrown into interplanetary space where it dissipates harmlessly.

Perspective on Future Research

The current U.S. program in solar physics includes participation in the Japanese Yohkoh mission, which carries a cluster of X-ray telescopes. It is an effective tool for studying high-temperature ionized flare plasmas. Particularly rapid progress is being made by mapping the overall structure of flares and their effects on the corona. A spectrometer on board reveals the temperature, density, and flow velocity of the hot plasmas. Finally, particle acceleration sites are being probed with hard X-ray and gamma-ray telescopes.

Yohkoh is effectively addressing many issues about the heating, location, and constitution of flare plasmas. Because its resolving power and that of the ground-based telescopes is not better than ~1000 km, we do not expect substantial progress on the fundamental trigger mechanisms and magnetic processes in flares. However, Yohkoh's eventual contribution to our understanding of the build-up and the distribution of flare energy may be very great indeed.

Over the past decade, coordinated ground- and space-based instruments successfully obtained flare spectra ranging from the gamma-ray to the radio-wave band. These observations have led to an understanding of how flares partition their energy among the various layers of the solar atmosphere. Yet, the processes that accelerate electrons and protons to relativistic energies remain elusive. Future progress in understanding particle acceleration will require images of the high-energy emissions with much higher spectral and spatial resolution. This will be the mission of NASA's High Energy Solar Physics (HESP) spacecraft. Current plans are to launch it before the next maximum of solar activity due in 2001. The central objective of the HESP mission will be to locate the particle acceleration sites and to place them in their magnetic field context.

Solar magnetic fields will be better detailed by a new generation of ground-based and balloon-borne optical and near-infrared magnetographs. Magnetic-field mapping has improved greatly in the past 5 years. The measurements are beginning to show some very interesting correlations between electric currents and flare sites. There are still many questions, however, about the adequacy of ground-based magnetographs to resolve the fundamental issues of magnetic energy release and build up. Limited spatial resolution and the intermittency of the observations are problems. There is an acute need for much sharper images than any ground-based telescope can make, and a NASA panel has devised a suite of balloon-borne and rocket experiments that will provide a low-cost look in the Mechanisms of Solar Variability program.

References

Foukal, P. V., *Solar Astrophysics*, Wiley, New York, 1990.
Lepping, R. P., J. A. Jones, and L. F. Burlaga, Magnetic field structure of interplanetary magnetic clouds at 1 AU, *J. Geophys. Res. 95*, 11,957, 1990.
Vrsnak, B, V. Ruzdjak, and B. Rompolt, Stability of prominences exposing helical-like patterns, *Solar Phys. 136*, 151, 1991.

David Rust
The Johns Hopkins University, Applied Physics Laboratory, Laurel, MD 20723

Solar Flare Particles

R. P. Lin

The Sun is the most powerful natural particle accelerator in our solar system, able to accelerate ions to energies of many GeV and electrons to hundreds of MeV. This acceleration occurs as a consequence of transient releases of energy in solar flares (*Rust*, this vol.) and/or coronal mass ejections (CMEs) (*Goldstein*, this vol.). Solar flares are explosions occurring near sunspots, regions of strong, $\sim 10^3$ Gauss, magnetic fields.

Traditionally, flares have been detected by ground-based optical observatories as brightenings in the emission of the hydrogen-alpha line. In the largest flares, as much as $\sim 10^{32}$ ergs is released in $\sim 10^3$ seconds. Energy is thought to be stored in the magnetic field and released through some type of instability. Because the number of sunspots waxes and wanes over an ~ 11-year cycle (*Hathaway*, this vol.), the frequency of occurrence of flares also varies with this solar activity cycle; the last solar maximum occurred in 1989–1990.

CMEs are transient ejections of material from the solar corona—the outer atmosphere of the Sun—into the interplanetary medium. On average, $\sim 5 \times 10^{15}$g of material are ejected per CME, traveling at speeds of ≤ 100 km/s to ~ 2000 km/s. About one-third of all CMEs are traveling fast enough to drive a collisionless shock wave. These shocks appear to be responsible for accelerating some of the solar energetic particles, mostly at energies below a few tens of MeV. Although many large solar flares are accompanied by a fast CME, only a few CMEs accelerate particles in the absence of flares while many flares accelerate particles without CMEs.

Some of the accelerated particles travel to the near-Earth interplanetary medium where they can be detected by spacecraft. The most energetic of these particles produce effects at the surface of the Earth. Following very large flare events in February and March 1942, S. E. Forbush first detected solar energetic particles (>6 GeV protons). Since then, observations with ground-based, balloon-borne, and spacecraft instrumentation have detected solar energetic protons, alpha particles, heavy nuclei, and electrons, from nearly thermal energies (a few keV) up to galactic cosmic-ray (*Jokipii*, this vol.) energies (~10 GeV).

Solar energetic particles can, on occasion, influence human activities (*Joselyn*, this vol.). In the most intense events, particles energetic enough to penetrate the walls of manned spacecraft can result in a harmful or even fatal radiation dose to astronauts. Such intense events also degrade components on unmanned spacecraft, such as solar panels and semiconductor electronics. Solar energetic particles can also penetrate deep into the atmosphere over the Earth's magnetic polar regions and produce increased ionization, lowering the ionosphere and disrupting radio communications.

Much of the solar particle acceleration occurs in the solar corona, where the accelerated ions and electrons interact with the ambient solar atmosphere to generate hard X-ray, gamma-ray, neutron, and radio emissions. These electromagnetic emissions provide information on the energetic particle populations at the Sun. In particular, analysis of the hard X-ray emission shows that for many solar flares the energy contained in electrons of tens of keV energy can amount to ~10-50% of the total energy released. Thus particle acceleration appears to be an integral part of the fundamental flare energy release process.

This ability to release energy impulsively and accelerate particles to high energies is shared by magnetized cosmic plasmas at many sites throughout the universe, ranging from magnetospheres to active galaxies. The basic physics of these processes can be studied best at the active Sun; the accelerated particles range up to cosmic-ray energies and the escaping particles can be sampled directly, while those interacting with the solar atmosphere can be observed via their electromagnetic emissions. Furthermore, the proximity of the Sun means the region where the flare energy release and particle acceleration takes place can be located and studied in detail. Such studies provide an understanding of these fundamental physical processes, and will eventually lead to a predictive capability useful for human endeavors in space.

Large Solar Energetic Particle Events

There are two main classes of solar energetic particle events in the interplanetary medium. The most intense and energetic type, large solar energetic particle (LSEP) events, produce a significant flux of >10 MeV protons (Figure 1). They usually occur after a large solar flare, and occasionally accelerate up to relativistic energies, as in the events reported by *Forbush* in 1946. Besides

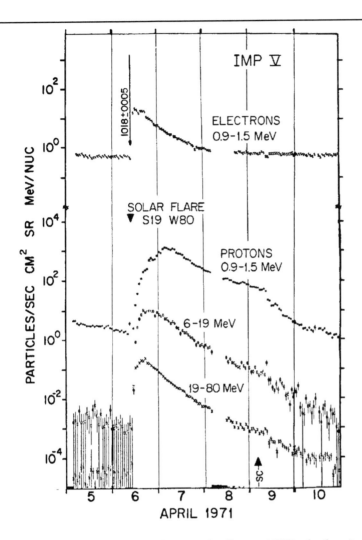

Figure 1. Example of a LSEP event from a solar flare at 80°W solar longitude. The passage of a shock wave is indicated by the sudden commencement (labeled SC at lower right) seen at the Earth.

energetic ions, electrons are also observed, but the fluxes of energetic protons dominate over electrons. Tens of LSEP events are detected per year near solar maximum.

The solar energetic particles propagate away from the Sun along the interplanetary magnetic field. Because of the very high electrical conductivity of the corona and solar wind, the solar magnetic field is essentially frozen into these plasmas. The radial outflow of the solar wind into interplanetary space

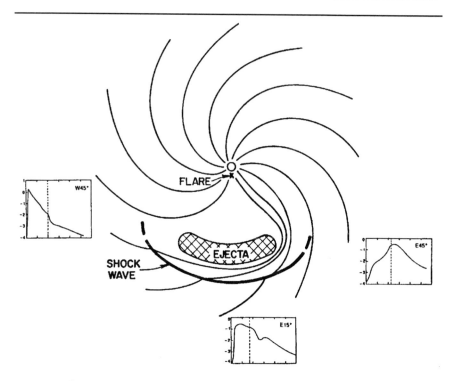

Figure 2. The interplanetary magnetic field is drawn out to an Archimedes spiral as shown in the upper half. In the lower half the rectangles show representative profiles of 20 MeV protons seen by observers when the source flare is at various solar longitudes, assuming a fast CME is propagating outward with a shock wave (heavy line) ahead. For each profile the dashed line indicates the time of passage of the CME shock wave. The vertical axis is the logarithm (base 10) of the proton flux, while each tick of the horizontal time scale is 1 day.

then draws the magnetic field out into an Archimedean spiral shape as the Sun rotates (Figure 2).

The magnetic field exerts a Lorentz force on the electrically charged accelerated particles, which leads to helical motion of the particles about the field direction. Thus solar energetic particles are guided along the Archimedean spiral field. For typical solar wind speeds of ~400 km/s, the foot of the field line leading to the Earth would be located at about 60°W solar longitude (defined relative to the Sun-Earth line as seen from the Earth) at the surface of the Sun (Figure 2). For the accelerated particles, then, flares located near this longitude are magnetically "well-connected."

For magnetically well-connected LSEP events, the energetic particle fluxes rise rapidly soon after a large solar flare (Figure 1). The first particles arrive within the travel time along the spiral field from the Sun. The fastest particles arrive earliest, in a manner usually consistent with particles of all energies being accelerated simultaneously at the Sun by the flare, and traveling the same distance along the spiral interplanetary magnetic field. For these LSEP events, the rapid rise to maximum is usually followed by a much slower decay, which has been attributed to scattering of the particles by irregularities in the interplanetary magnetic field, and/or to prolonged acceleration. Occasionally there are events where the particles appear to propagate essentially "scatter-free."

The plasma of the interplanetary medium is so tenuous and collisions so infrequent that once the energetic particles leave the solar corona they do not change their ionization state. Measurements of the ionization states of LSEP particles show that they are typical of a 1–2 10^6 K plasma. This suggests that the LSEP particles come from the quiescent solar corona and not from the hot, $\geq 10^7$ K, solar flare plasma or from the solar photosphere or chromosphere where temperatures are less than 10^5 K. The elemental composition of the LSEP particles appears to reflect typical coronal abundances, and in fact, LSEP measurements are one of the best sources of information on the solar corona's elemental abundances.

LSEP events are often observed from solar flares far from the well-connected magnetic foot-point of the interplanetary field line to the observer. Studies indicate that LSEP particles may be spread over about ±10° in solar longitude away from the flare site. It was initially thought that some sort of cross-magnetic field diffusion near the Sun could account for the spreading of the particles across solar longitude, but the lack of compositional variation with longitude and the rapid onsets for LSEP events even for flares far from the connection longitude, suggests that the acceleration is widespread. The observed coronal ionization states and elemental composition indicate that some of these LSEP particles, at energies below tens of MeV energies, are accelerated by CME shock waves as they propagate over a wide longitude range of the solar corona [*Lin*, 1987].

An increase in the energetic ion fluxes, particular at low energies, is often observed when a fast CME shock passes the spacecraft, indicating that the shock can continue to accelerate particles even in the interplanetary medium near 1 AU. At late times in LSEP events, the shape of the time profile of the particle fluxes varies with solar longitude of the flare (Figure 2) in a way that is consistent with distortion of the interplanetary magnetic structure by the CME, and with continued acceleration by the shock wave as the CME and shock propagate outward through the interplanetary medium [*Cane et al.*, 1988]. Above several tens of MeV energies, however, interplanetary shock acceleration effects are rarely detected. At times of intense solar activity, multiple solar flares and CMEs from a single active region can continually accel-

Table 1. Solar Energetic Particle Event Characteristics

Characteristic	Large Solar Energetic Particle (LSEP) Events	Non-relativistic Electron - ^{3}He-rich Events
Dominant Particle Species	$\geq$ MeV protons	~2-100 KeV electrons
Electron to Proton Ratio	small	large
^{3}He/^{4}He Ratio	"normal" solar (~5x 10^{-4})	~0.1 - 1 (~10^2 to $\geq 10^3$ times normal)
Heavy Nuclei	"normal" solar	enhanced abundances of Fe, Mg, Si, S
Ionization States	typical of 1 - 2 x 10^6 K normal corona	highly stripped (e.g. Fe^{20}) typical of ~10^7 K plasma
Extent in Solar Longitude	$\geq 100°$	tens of degrees
Event Rate (at solar maximum)	tens per year	$\geq 10^3$ per year
Flare Association	large solar flare (but sometimes missing)	mostly small flares but often no flare
Solar Soft X-ray Burst	gradual, > 10 minute duration	impulsive, < 10 minute duration
Interplanetary Association	coronal mass ejection with fast shock	interplanetary type III radio burst

erate particles to form long-lived "super-events" about 40 days long that fill the heliosphere with solar energetic particles at energies up to tens of MeV [*Dröge et al.*, 1992].

Since LSEP events and fast CMEs are typically accompanied by flare soft X-ray emission of relatively long duration, with e-folding decay times more than tens of minutes, LSEPs are also called "gradual" events. The characteristics of LSEP events are listed in Table 1.

Solar Nonrelativistic Electron–^{3}He-Rich Events

The second type of solar energetic particle event is dominated by electrons (Figure 3). The first unambiguous detections of solar energetic electrons were made in 1965 by spacecraft in the interplanetary medium. These were electrons of tens of keV energies, and it soon became apparent that such nonrelativistic electron events occur far more frequently, ~10^3 events per year at solar maximum, than the LSEP events.

More recent measurements at energies down to 2 KeV showed that the electron spectrum usually continues to monotonically increase down to 2 KeV, and many events were only detected below ~10 KeV [*Lin*, 1985]. At these low energies the electrons must have been accelerated high in the solar corona, or they would be lost to Coulomb collisions before escaping the Sun. As the electrons travel through the interplanetary plasma they excite plasma waves, which in turn produce solar emission called type III radio bursts. Such bursts are observed in the interplanetary medium for nearly every nonrelativistic electron event.

At first, nonrelativistic electron events appeared not to be accompanied by detectable solar energetic ion fluxes; in contrast to LSEP events, the energetic electron to proton ratio is generally very large. In 1970, ^{3}He-rich solar particle events were first discovered. These are weak events dominated by low energy (<1 Mev/nucleon) ions.

Often protons as well as ^{4}He were also observed to be underabundant. However, as experiments improved in sensitivity, ^{3}He-rich emission was observed to accompany nearly all of the more intense nonrelativistic electron events (Figure 3). Thus some ^{3}He-rich ion emission may accompany all nonrelativistic electron events.

The nonrelativistic electron-^{3}He events are typically associated with small flares or subflares or lack a flare entirely, and there is no association with CMEs. The electrons and ^{3}He nuclei often propagate nearly scatter-free through the interplanetary medium and quickly escape the inner solar system, leading to relatively rapid decay of the fluxes. The acceleration of the ^{3}He ions and electrons occurs close to the same time. The associated flares generally have impulsive, short soft X-ray bursts, so these are also called

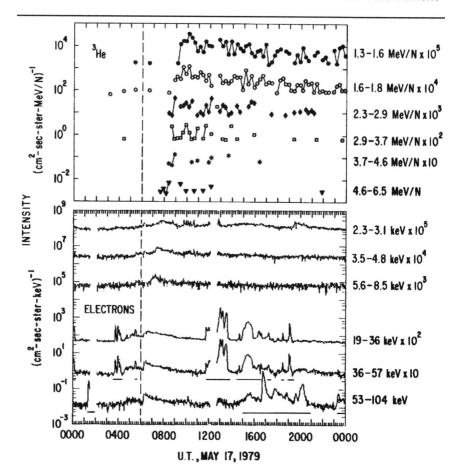

Figure 3. Nonrelativistic electron-<^{3}He rich event. The lower panel shows an electron event beginning at ~0620 UT at 53–104 keV. Velocity dispersion is clearly evident with the lowest energy (2.3–3.1 keV) electrons beginning ~0730 UT. The horizontal lines indicate times of contamination by terrestrial upstream particles or solar X-rays. The upper panel shows the associated ^{3}He event, which also exhibits velocity dispersion. In this event the ^{3}He/^{4}He ratio is >10 and no solar flare was observed. The vertical dashed line indicates the time of the solar type III radio burst.

"impulsive" events. The flares are magnetically well-connected near 60°W longitude, with a spread of only a few tens of degrees in longitude, indicating that the acceleration process is probably localized near the flare.

The elemental composition in ^{3}He-rich events shows substantial enhancement in the abundances of heavy nuclei; Mg, Si, S, Ne, Fe, but not O, C, N

[*Reames et al.*, 1994]. The ionization states are much higher than for LSEPs; Fe is detected with an average charge state of around +20 for ^{3}He-rich events compared to around +13 for LSEPs, suggesting that either the source has a temperature of $\geq 10^7$ K, comparable to the temperature of the hot flare plasma, or perhaps that the accelerated ions were stripped of some of their bound electrons.

Nonrelativistic electron-^{3}He-rich events (Table 1) thus comprise a class of solar particle acceleration events distinct from LSEPs. The enormous enhancement of ^{3}He suggests some sort of resonant acceleration process [e.g., *Temerin and Roth*, 1992].

Solar Energetic Particles Observed at the Sun

Much of the particle acceleration occurs close to the Sun, where the fast electrons generate hard X-rays through bremsstrahlung collisions, and the fast (≥ 10 MeV) nuclei generate gamma-ray lines through nuclear collisions with the ambient solar atmosphere. The fast electrons can also produce microwave radio emission through the gyro-synchrotron process in the strong magnetic fields, and decimetric, metric, and decametric radio emission through the electron beam-plasma process.

Figure 4 shows hard X-ray and gamma-ray emission from a large solar flare. The emission lasts about a minute, and consists of a series of ~5-10 s individual peaks. The hard X-ray emission is generated by fast electrons, and the two profiles are essentially simultaneous at all energies. The 4.1–6.4 MeV gamma-ray emission, produced predominantly by nuclear collisions of ≥ 10 MeV ions, is very similar to the hard X-rays, but with a delay of about 2 s. Thus electrons and ions appear to be accelerated to high energies nearly simultaneously on short time scales. Gamma-ray and neutron emission at energies of tens to hundreds of MeV also have been observed from large flares. These emissions have similar time scales, indicating that ions and electrons of hundreds of MeV to GeV energies are also accelerated rapidly.

Hard X-ray emission above 20 keV is observed for most solar flares, even small ones, and perhaps is produced in all flares. Since the bremsstrahlung collision process is well understood, detailed hard X-ray observations provide quantitative estimates of the numbers and energies of fast electrons at the Sun. The fast electrons lose most of their energy to Coulomb collisions with the thermal electrons in the ambient medium, with only a small fraction going to photon-producing bremsstrahlung collisions, provided that the fast electrons are much more energetic than the ambient thermal electrons. Then, for many flares, ~10-50% of the total energy released must be initially in accelerated ≥ 20 keV electrons.

Much of the observed flare phenomena—optical brightenings, evaporative

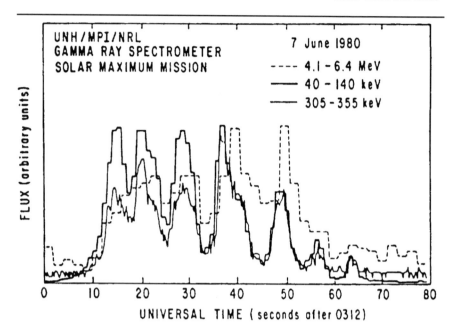

Figure 4. Hard X-ray and gamma-ray burst for the June 7, 1980, solar flare. The 40–140 and 305–355 keV channels show bremsstrahlung emission from energetic electrons. The 4.1–6.4 MeV channel is dominated by nuclear gamma-ray line emission produced by >10 MeV ions.

mass motions, production of hot ~10^7 K flare plasma—can be attributed to the heating of the solar atmosphere by the accelerated electrons! Little is known, however, about the electron acceleration mechanism, although there is some evidence for strong electric fields playing a role. Even less is known about the acceleration of ions [*Mandzhavidze and Ramaty*, 1993]. Another important question is, how are the energetic particle populations at the Sun related to the ones detected in the interplanetary medium? Many interplanetary nonrelativistic electron events, especially in the 2–10 keV range, are not associated with solar flare or hard X-ray burst and vice versa. Similarly some LSEP events have little or no associated gamma-ray emission and vice versa. The nuclear gamma-ray line spectrum can provide elemental composition information about the accelerated ions and the ambient solar atmosphere. In the single gamma-ray flare event that has been analyzed in detail, an enhancement of the accelerated heavy ions, Fe in particular, relative to normal coronal abundances, appears necessary to best fit the observations. Furthermore, the ratio of energetic electrons to energetic protons for nuclear gamma-ray flares is larger than observed for LSEP events. Thus nuclear gamma-ray flares may be more similar to nonrelativistic electron-^{3}He-rich events than to LSEP events (see Table 1). The soft X-ray bursts for these flares, however, can be either impulsive or gradual.

The particle acceleration and energy release mechanisms for flares still remain mysteries. Future spacecraft missions, such as the High Energy Solar Imager (HESI) and Advanced Composition Explorer (ACE), should provide some of the key observations required for understanding of the physics of these fundamental processes.

References

Cane, H. V., D. V. Reames, and T. T. von Rosenvinge, The role of interplanetary shocks in the longitude distribution of solar energetic particles, *J. Geophys. Res.*, *93*, 9555, 1988.

Dröge, W., R. Müller-Mellin, and E. W. Cliver, Superevents: Their origin and propagation through the heliosphere from 0.3 to 35 AU, *Astrophys. J.*, *387*, L97, 1992.

Lin, R. P., Energetic solar electrons in the interplanetary medium, *Sol. Phys.*, *100*, 537, 1985.

Lin, R. P., Particle acceleration and propagation, *Rev. Geophys.*, *25*, 676, 1987.

Mandzhavidze, N., and R. Ramaty, Particle Acceleration in solar flares, *Nucl. Phys. B*, *33*, 141, 1993.

Reames, D. V., J. P. Meyer, and T. T. von Rosenvinge, Energetic-particle abundances in impulsive solar flare events, *Astrophys. J.*, *90*, 649, 1994.

Temerin, M., and I. Roth, The production of ^{3}He and heavy ion enrichments in ^{3}He-rich flares by electromagnetic hydrogen cyclotron waves, *Astrophys. J.*, *391*, L105, 1992.

R. P. Lin

Physics Department and Space Sciences Laboratory, University of California, Berkeley, CA 94720

Solar Irradiance Variations and Climate

Peter Foukal

In 1838, the French physicist Claude Pouillet published the first measurement of the Sun's total light and heat input to the Earth. He described his new instrument—the pyrheliometer—and the corrections he made for attenuation of solar light in the Parisian atmosphere. Similar measurements were carried out by the English astronomer Sir John Herschel, working at about the same time at the Cape of Good Hope.

Pouillet used the value of 1.76 calories cm^{-2} min^{-1} that he obtained for his "solar constant" to calculate that sunlight at the Earth's surface was powerful enough to melt a global ice layer 31 m deep in 1 year. He also pointed out that this value implied a prodigious solar power output whose source deserved consideration, and he attempted to determine the temperature of the solar surface emitting this vast flux of heat and light.

We no longer wonder at the magnitude of the Sun's power output per square meter—sufficient to power half a dozen Atlantic liners at their utmost speed, night and day—as 19th-century astronomers enjoyed pointing out. The source of power in a star like the Sun is also less enigmatic to us than it was to Victorian astronomers, although satisfactory closure of the neutrino problem may yet require modification of our understanding of energy generation, mixing, and convection in the solar interior.

The frontier in studies of solar irradiance has gradually moved toward variability of the Sun's total output, and also of its outputs in the infrared, ultraviolet, and extreme ultraviolet spectral regions. The last two wavelength regions are largely inaccessible to ground-based measurement, due to strong absorption in the Earth's atmosphere. The motivation for these studies has progressed beyond the enthusiasm of Pouillet and his contemporaries for applying the newly discovered laws of thermodynamics to a star and to the Earth's atmosphere.

Population pressures in 1994 now bring us to a very practical concern about the role of a variable Sun in changes of climate and ozone, and in possible delicate couplings between the Earth's outermost atmosphere and its biosphere.

Total Irradiance Variations

In the 1880s, Samuel Langley and Charles Abbot measured the spectral distribution of the solar output in the infrared and wondered about how its variability might influence climate through its absorption in the strong terrestrial water vapor bands. This work eventually led to an epic series of measurements designed to detect small variations in the solar constant, which lasted until 1955. This program included a 32-year run of daily measurements from several mountain stations in North and South America and Africa.

Analysis of the data showed that if any long-term variations occurred, they were below the 1% reproducibility level of the measurements over timescales of climatological interest. This was a significant result, since no comparable constraint on solar output could be placed from other sources of information on the solar atmosphere and interior. Unfortunately, Abbot's insistence on doubtful correlations between variations in solar constant and weather eventually gave the topic a bad reputation. This unfortunate situation persisted until more precise radiometry from space became available in the late 1970s.

We owe our present understanding of variations in the solar total irradiance, S, to analysis of the daily data obtained since 1978 by pyrheliometers flown on the Nimbus-7 and Solar Maximum Mission (SMM) satellites. More coarsely sampled data obtained from the Earth Radiation Budget Satellite (ERBS) have been very useful in checking the long-term trends. The overall picture of ΔS derived from these three data sets is shown in Figure 1.

Note the lower value of S measured in the period of solar activity minimum around 1986, compared to the two maxima around 1980 and 1991. This shows that the Sun is brighter at high activity levels, not dimmer as we might expect given the greater area coverage by dark sunspots near peaks of the sunspot cycle. The amplitude of this 11-year variation in S is about 0.1%. The exact value is somewhat uncertain because the reality of the large peak in the Nimbus-7 radiometry in 1979 remains controversial.

Another significant feature is the high-frequency variation of the irradiance signal. Its amplitude is about 0.2% during the higher-activity phase of the solar cycle, but much lower around solar activity minimum. Comparison of the measured variation with the ΔS calculated from the areas and photometric contrast of dark sunspots shows that much of the irradiance fluctua-

Solar Irradiance from Three Satellites

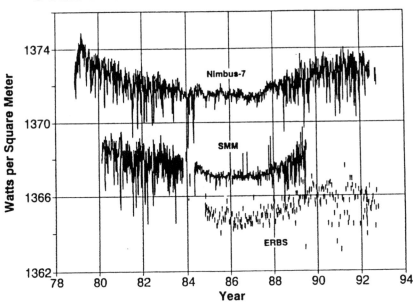

Figure 1. Daily irradiance values are plotted for three independent satellite sensors: the Nimbus-7 ERB (November 1978 to September 1992), SMM ACRIM (February 1980 to June 1989), and ERBS solar sensor (October 1984 to October 1992). The vertical displacements between the measurements are caused by uncertainties in the three absolute calibration. (Courtesy of L. Kyle)

tion consists of dips lasting a few days, as large spot groups transit the solar disc. Most of the remaining high-frequency variance is caused by disc passage of faculae, which are extended, bright magnetic structures often associated with sunspots. Examples of sunspots and faculae are shown in Figure 2. Figure 3 gives an example of the good agreement obtained between calculated and observed time series of daily Δ*S*.

The brightening of the Sun measured around the maxima of sunspot cycles 21 and 22 appears to be caused by the dominant contribution of bright magnetic faculae to the disc-integrated irradiance signal. That is, while spots have greater photometric contrast than faculae, the faculae cover a larger fraction of the solar surface, including a substantial fraction of the area outside of the active regions where spots are concentrated (Figure 2).

The plausibility of this picture of solar total irradiance variation can be judged further from Figure 4, where the solid curve shows the Active Cavity Radiometer Irradiance Monitor (ACRIM) radiometry from the SMM satellite

Figure 2. Bright faculae in an active region and in the network imaged together with small sunspots near the Sun's limb. (Courtesy of G. Chapman)

between 1980 and 1990. The dashed curve represents the prediction from an empirical model whose inputs are the daily coordinates of spots, their projected areas and broadband photometric contrast, and the daily value of a proxy indicator of facular area.

The variations seen in the simple model agree with variations in the data over this time period, when the accuracy of the radiometry was highest. To achieve this agreement, it was not sufficient to consider only the contribution of faculae in the active regions. The variations on the 11-year timescale are only reproduced if we include in the model the disc-integrated contribution of tiny bright faculae distributed over the entire Sun, in the so-called magnetic network.

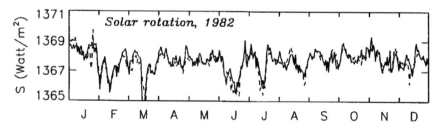

Figure 3. Plots of irradiance variations measured by the ACRIM experiment (solid) and of irradiance values calculated from daily sunspot and facular proxy data (dashed). (Courtesy of J. Lean)

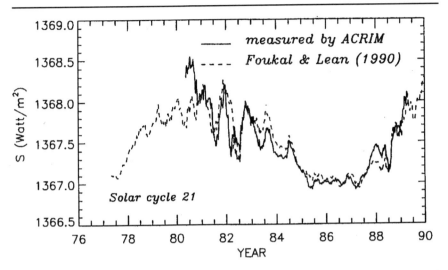

Figure 4. Irradiance values measured by the ACRIM experiment between 1980 and 1990 (solid), compared with a model of irradiance variations based upon sunspot data and a proxy indicator of faculae (dashed). (Courtesy of J. Lean)

Information on the changes in area and intensity of the magnetic network over the solar cycle must be obtained from its high-contrast radiations in microwaves and in line radiations formed in higher layers of the solar atmosphere overlying the photosphere. To make the model completely convincing, one should measure variations in the area and brightness of the network directly in the photosphere, where 99% of the total solar luminosity originates. This is difficult because the network's photometric contrast in these layers is low. Such photometry is now being planned using the precision solar photometric telescopes (PSPTs) being developed at the National Solar Observatory. The PSPT project is an element of the Radiative Inputs of the Sun to Earth (RISE) Program supported by the National Science Foundation.

An empirical model can be used to estimate the variability of S caused by solar surface magnetism back to 1874, when reliable daily measurements of spots (and faculae) began at the Royal Greenwich Observatory. Such a model indicates that the amplitude of 11-year variations is typically about 0.05% and that the irradiance variation seen in cycle 21, which maximized around 1981, was the largest of any cycle since 1874.

The simplified treatments of spot, facular, and network contributions to total irradiance made in this model are worth checking, because the model's conclusions have been widely used to evaluate the possible solar contribution to global warming since the 1850s. However, it is unlikely that any of the corrections will influence the result that irradiance variations generated by photospheric magnetic structures over the past 150 years seem relatively small, with a peak excursion of less than 0.1%.

When conclusions regarding climate change are drawn from a solar irradiance model of this kind, we must remember that the Sun's luminosity could vary on timescales longer than the 11-year cycle, at amplitude levels not controlled by photospheric magnetism.

For instance, the dominant scales of convective heat transport to the solar photosphere are still quite unknown. If large-scale convection cells prove to be important, substantial changes in heat transport efficiency might be expected, judging from the random fluctuations seen in dynamical models of solar convection. Convincing evidence for such long-term solar luminosity variations can come only from continued long-term monitoring of the Sun's total irradiance from a sequence of satellites. But valuable indirect evidence might come more quickly through study of the dynamics of solar convection and from photometry of other Sun-like stars.

Changes in Solar Convective Heat Flow

New insights into heat flow on the Sun may come from photometric study of photospheric brightness inhomogeneities other than those caused directly by the magnetic fields of spots, faculae, and network. Such measurements are important, but have not uncovered convincing heat flow perturbations on spatial scales larger than the well-known photospheric granulation, which is seen as a low-contrast mottling in Figure 2.

Another path is through detailed study of the irradiance signature of the spots and faculae themselves. The darkness of spots is widely believed to be caused by the local inhibition of solar convection by intense, radially oriented magnetic fields. If this is correct, the heat not radiated from a dark spot could be blocked and mainly stored as thermal and potential energy in the convection zone.

This interpretation is suggested by the results of modeling obstructions to solar heat flow. Here the convective processes are parameterized as a turbulent diffusion, and the relevant radiative boundary condition is used at the photosphere. Whatever the details of solar convection may be, the radiative leak is slow enough that the dips in the radiometry are easily explained. Valuable information on the spatial scale of the most effective convective modes may be extracted as better models of convection around sunspots are compared with radiometric and photometric measurements.

The most widely accepted explanation of facular brightness builds on similar ideas. The main difference may be the much smaller cross-sectional area of the facular magnetic flux tube. In both cases, magnetohydrostatic equilibrium leads to much lower plasma pressure inside the flux tube than outside. This lower opacity enables intense radiation from the hot interior wall of the flux tube to escape, causing it to act as a thermal leak.

The relative importance of this thermal leak should scale as the circumference of the flux tube, while the magnetic inhibition of convection in the same flux tube scales as its area. Thus the small-diameter facula could reduce the net local thermal impedance, producing a bright structure.

Observations of Sun-Like Stars

Further evidence on the Sun's luminosity variation comes from photometry of other stars similar in mass to the Sun. Studies indicate that stars younger and more magnetically active than the Sun become fainter with increasing activity level during their activity cycle; not brighter, like the Sun. Analysis of their photometric light curves indicates that very large spots form on more active stars, and their darkening of the star's surface overwhelms the brightening caused by the faculae.

This result agrees with the finding from solar studies that the ratio of facular to spot area decreases at the highest activity levels. This suggests that the Sun might well become dimmer, rather than brighter, if its activity level were to increase. A more active Sun would also be more variable, since faculae would be much less effective in compensating the luminosity decrease of the very large spots. However, evidence indicates that the Sun is now about as active as at any time in the past several millennia, so an increase of irradiance variation driven by photospheric magnetic activity seems unlikely in the present epoch.

Further study of the results from stellar photometry is required to determine whether stars exist which exhibit large luminosity variations that cannot be ascribed to photospheric magnetic activity. Discovery of such stars would be an important milestone for climate studies because the imminent occurrence of such possibly large-amplitude luminosity variations on the Sun would be much harder to rule out.

Ultraviolet and Extreme Ultraviolet Flux Variations

Indications of the Sun's variable EUV and UV outputs were obtained in the 1930s from correlations between shortwave radio fadeouts and solar flares [*Rust*, this vol.], and preparations were made in Germany during WW II for measurements using UV-sensitive crystals mounted in a spectrometer on a V-2 rocket. But the end of the war intervened, and the first UV spectra showing the Sun's radiations below the atmospheric cutoff around 300 nm were obtained in 1946 by the Naval Research Laboratory group, using captured V-2s. By 1949, comparison of the EUV and X-ray rocket data flown at varying levels of solar activity confirmed that the solar outputs in the EUV and soft X-ray ranges were highly variable.

We now know that in the ultraviolet wavelength range of greatest importance to ozone studies, namely between ~160 and 320 nm, the short-term variation of spectral irradiance decreases from about 10% to less than 1%, respectively. These are mainly continuum radiations, emitted predominantly in the relatively cool photospheric and chromospheric layers of the Sun's atmosphere. The UV flux variations in this wavelength range are caused mainly by evolution and rotation of the same magnetic faculae, sunspots, and network that cause the changes in solar luminosity described above. Direct information on changes in the faculae and network back to 1915 has recently become available through digitization of daily, full-disc solar images obtained at Mt. Wilson Observatory between 1915 and 1984. This digitization effort is another element of the RISE program.

Below about 150 nm, the dominant contribution comes from line radiations emitted in the overlying coronal plasmas of temperatures up to several million degrees. With some notable exceptions, such as Lyman-alpha at λ 121.6 nm, these radiations are optically thin, and their intensity dependence upon the square of the highly fluctuating plasma density makes them more variable than the increasingly optically thick emissions of cooler atmospheric layers. Short-term variability of the EUV line irradiances at the 25% level is common, and flares can increase irradiances by a factor of 10 or more in this range.

Empirical models are less accurate in the EUV, because the structures observed in the strong resonance lines of ionized heavy elements such as Ne, Mg, Si, and Fe, consist of complex magnetic loop systems connecting between active regions and the quiet Sun. The intensities of these loops vary so much that it is problematic to attempt models based on a tractable set of photometric variables, such as spot and facular area, and unique values of photometric contrast, observable from the ground. There may be no practical alternative to space-based monitoring of the EUV spectral irradiance.

The Sun's UV and EUV variability could provide interesting information on nonthermal heating mechanisms in the Sun's atmosphere. We still possess only a sketchy understanding of the wave or electric current dissipation mechanisms responsible for producing the Sun's high levels of UV and EUV radiations. Measurement of a systematic solar-cycle dependence of the brightness of UV faculae or network could provide insight into similar variation of the heating function. Such information would pose interesting new constraints on this classic problem of stellar physics.

In this connection, we note that the brightness of sunspot umbrae has been shown to increase systematically over the solar cycle at the 15% level. This unexpected result has been measured over two complete 11-year cycles

[*Hathaway*, this vol.]. So far, it is a mystery how a spot's magnetic refrigeration mechanism could respond to the phase of the solar cycle.

The amplitude of spectral irradiance variation over the 11-year cycle is the issue of greatest practical importance in the study of UV and EUV irradiance variations. The answer to this key question will determine the solar influences on ozone and upper atmospheric density, with its effect on satellite drag. The data released from the SUSIM and SOLSTICE experiments on the Upper Atmosphere Research Satellite (UARS) still span only a small fraction of the 11-year cycle. But the factor of 2 variation in the EUV below about 120 nm indicated by previous data seems consistent with these newest measurements through the end of 1992, with a decrease to less than 10% longward of 200 nm.

It is extremely important to aeronomy and upper atmospheric physics that UARS measurements continue at least through the next solar activity minimum. Preferably, they should extend as far into the next cycle (23) as is technically feasible. They offer the space physics community, and NASA, a rare opportunity to answer a well-posed question of broad practical importance.

Beyond UARS, the ability to interpret past climate variations and to foresee those of the future will require a long radiometric sample of total solar irradiance behavior. Its length must be comparable to the 50-year timescale of global temperature variations associated with the Little Ice Age, and also with the more recent period of global warming.

Maintaining a radiometric scale in space to within 0.02% or better for 50 years is technically challenging. This level of radiometric accuracy and stability can now be attained in laboratories using the new technique of cryogenic radiometry, but its achievement in space may be more difficult because of potential contamination problems. Nevertheless, ground-based calibration of flight radiometers should significantly increase the accuracy of long-term irradiance measurement programs. Progress will require substantial resources for such a long-term program, whose full benefit will only be realized by a future generation of space and climate physicists. Mounting concerns over global change, driven by increasing population pressure, are spurring our community and NASA to confront this important challenge.

Peter Foukal
Cambridge Research and Instruments, Inc., 21 Erie St., Cambridge, MA 02139

The Solar Dynamo

David H. Hathaway

T he solar dynamo is the process by which the Sun's magnetic field is generated through the interaction of the field with convection and rotation. In this, it is kin to planetary dynamos and other stellar dynamos. Although the precise mechanism by which the Sun generates its field remains poorly understood despite decades of theoretical and observational work, recent advances suggest that solutions to this solar dynamo problem may be forthcoming.

Two basic processes are involved in dynamo activity. When the fluid stresses dominate the magnetic stresses (high plasma $\beta = 8\pi p / B^2$), shear flows can stretch magnetic field lines in the direction of the shear (the "ω effect") and helical flows can lift and twist field lines into orthogonal planes (the "α effect"). These two processes can be active anywhere in the solar convection zone but with different results depending upon their relative strengths and signs. Little is known about how and where these processes occur. Other processes, such as magnetic diffusion and the effects of the fine scale structure of the solar magnetic field, pose additional problems.

Observed Behavior

Observations of sunspots and solar activity since the mid 17th century show that solar activity associated with the Sun's magnetic field waxes and wanes with an approximate 11-year cycle. The number of sunspots and the area they cover rise rapidly from minima near zero to maxima 3 to 4 years later. The decline from maximum then progresses more slowly over the remaining years of each cycle. Most measures of solar activity show this asymmetric rise and decline but exhibit substantial variations from one cycle to the next. During the Maunder Minimum of 1645 to 1715, the

DAILY SUNSPOT AREA AVERAGED OVER INDIVIDUAL SOLAR ROTATIONS

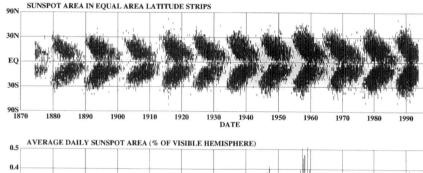

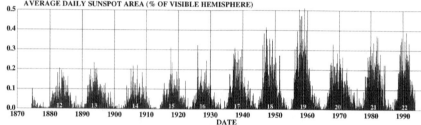

Figure 1. Sunspot areas and positions from 1874 to 1994. (upper panel) The latitudinal positions of sunspots are marked for each rotation of the Sun. This illustrates the equatorward movement of the active latitude band over each solar cycle. (lower panel) The average daily sunspot area, expressed as a percentage of the area of the visible hemisphere, is plotted for each rotation of the Sun. This illustrates the 11-year sunspot cycles and shows the cycle-to-cycle and rotation-to-rotation variations in total sunspot area.

sunspot cycle seems to have ceased entirely. This nonlinear and sometimes chaotic behavior suggests that the dynamo is not a simple wave or oscillatory phenomenon.

Sunspots do not appear randomly over the surface of the Sun but are concentrated in two latitude bands. This is best illustrated by a butterfly diagram (Figure 1), which marks the latitudes at which sunspots appear for each 27-day rotation of the Sun from May 1874 to June 1994. At the beginning of a cycle, sunspots appear only in the midlatitudes near 30°. As the cycle progresses, the latitude bands widen and move toward the equator where they disappear at the next minimum. This equatorward movement of the activity bands, known as Spörer's Law, suggests the presence of an underlying flow or wavelike propagation for the source of the activity. Sunspots tend to occur in groups that are strung out along a mostly east-west line. Spots within a group precede or follow in reference to the Sun's rotation. These groups usually are tilted so that the preceding spots are closer to the equator than the following spots (Joy's Law).

Direct measurements of the Sun's magnetic field began in 1908 and show that sunspots are sites of intense magnetic fields that are cooler and therefore dimmer than their surroundings. Early magnetic measurements revealed Hale's Polarity Laws: The preceding spots have one polarity while the following spots are of opposite polarity; the polarity of the preceding spots in one hemisphere is opposite the polarity of the preceding spots in the other hemisphere; and the polarities reverse from one 11-year sunspot cycle to the next to produce a 22-year cycle for magnetic activity.

Observations of weak magnetic fields provide additional details about the dynamo. After the strong fields erupt through the surface to form sunspots and active regions, the field elements spread out across the surface of the Sun. The field becomes concentrated in the network of downdrafts that outline the supergranule convection cells. As the supergranulation pattern evolves, the magnetic network evolves as well. The weak field observations reveal a slow poleward migration of these elements and the presence of weak polar fields that reverse polarity at about the time of solar maximum. The actual field has a fibril nature. In weak field regions the field is concentrated in small flux tubes that are surrounded by field-free regions.

Fluid Dynamic Properties

Models of the solar dynamo involve fluid motions within or adjacent to the solar convection zone that comprises the outer 30% of the Sun. These models should be consistent with the observed motions. Doppler velocity measurements and feature tracking provide information on flows at or near the top of the convection zone, while helioseismology provides information on flows in the interior.

The relevant flows include rotation, differential rotation—which is a variation in rotation rate with latitude and radius—meridional circulations, and convection. The Sun rotates about every 27 days but the equatorial regions rotate more rapidly (24 days) and the polar regions rotate more slowly (>30 days). Small variations on this rotation profile occur over the course of the solar cycle. The rotation tends to be slower near sunspot maximum and in the hemisphere with more spots, and slower in cycles with more spots. Rotating streams are observed in conjunction with the sunspots. These streams move toward the equator like the sunspots but appear to start earlier and at higher latitudes. The meridional flows at the surface are weak and thus difficult to measure but most observations indicate the presence of a flow of ~10–20 m/s from the equator toward the poles. The convective flows exhibit a wide range of size and behavior from granules that are 1000 km across and last about 20 minutes, to supergranules that are more than 40,000 km across and last for days.

Helioseismology probes the interior of the Sun by measuring the characteristics of sound waves produced by the turbulent convective flows. These waves, or *p*-modes, are trapped inside the Sun by the rapid change in density at the surface and the increasing sound speed deeper inside the Sun. The internal rotation can be measured by comparing the frequencies of waves moving prograde and retrograde for *p*-modes that sample different latitudes and depths. These observations of the internal rotation show that the observed surface rate extends inward through the convection zone along radial lines for each latitude. At the base of the convection zone the latitudinal differential rotation disappears and the rotation becomes more uniform. Radial gradients in the rotation rate occur primarily at the bottom of the convection zone with only very weak radial gradients throughout the bulk of the zone itself.

Long observation shows that other stars have activity cycles much like the Sun's. For a given stellar type, the level of activity increases with rotation rate. Cyclic behavior is found primarily in slow rotators like the Sun and amongst these, a quarter to a third appeared to be inactive during the years of observation.

Dynamo Theories

How are dynamos that produce similar behavior constructed? Early dynamo work showed what wouldn't work, but nonaxisymmetric flows provide the key for unlocking a variety of possible dynamos. In one branch of dynamo theory—mean-field electrodynamics—these nonaxisymmetric flows are represented by an average of their dynamical properties. Another branch—large eddy simulation—directly simulates the largest of these flows. Each approach has its own advantages, but neither produces a model in agreement with all the observations.

The basic equation of dynamo theory is the magnetic induction equation constructed from Maxwell's equations and Ohm's law:

$$\frac{\partial \mathbf{B}}{\partial t} = \nabla \times (\mathbf{v} \times \mathbf{B}) + \eta \nabla^2 \mathbf{B}, \tag{1}$$

where $\mathbf{B}$ is the magnetic induction, $\mathbf{v}$ is the fluid velocity, and η is the magnetic diffusivity. In mean-field electrodynamics, both the velocity and the magnetic induction are separated into mean and fluctuating parts. An average of the induction equation gives the mean-field equation that contains a new induction term given by the average of the cross product of the fluctuating velocity and magnetic induction. To first order, this term is proportional to the magnetic induction and its curl so that

$$\overline{\mathbf{y}' \times \mathbf{B}'} \approx \alpha \overline{\mathbf{B}} - \beta \nabla \times \overline{\mathbf{B}} \qquad (2)$$

where the primes denote fluctuating quantities, the overbar denotes an average, the constant α is proportional to the helicity in the fluctuation velocity field, and the constant β is proportional to the eddy diffusivity. Using spherical polar coordinates (r, θ, ϕ), equation (1) can then be written in terms of the mean toroidal (azimuthal) component of the magnetic induction, B_ϕ, and the poloidal (radial/meridional) component, $\mathbf{B}p=(Br, B_\phi)=\nabla \times A_\phi$, where A_ϕ is the vector potential. This gives a pair of coupled equations with

$$\frac{\partial B_\phi}{\partial t} + \left(\mathbf{U}_p \cdot \nabla \right) B_\phi = \left(B_p \cdot \nabla \right) U_\phi$$

$$\qquad (3)$$

$$+ \alpha \nabla \times \mathbf{B}_p + \beta \nabla^2 B_\phi$$

and

$$\frac{\partial A_\phi}{\partial t} + \left(\mathbf{U}_p \cdot \nabla \right) A_\phi = \alpha B_\phi + \beta \nabla^2 A_\phi \qquad (4)$$

where $\mathbf{U}$ is the mean fluid velocity consisting of meridional flow, $\mathbf{U}p$, and differential rotation U_ϕ. Neglecting for the moment the meridional flow, equation (4) shows us that the poloidal field is produced by the α–effect, in which the toroidal field is lifted and twisted by the nonaxisymmetric helical motions. Equation (3) shows us that the toroidal field is produced by both the α-effect and by the ω-effect, in which the poloidal field is stretched out by the differential rotation. The relative strength of these different terms determines the nature of the resulting dynamo.

If the differential rotation is much weaker than the α–effect, then the ω-effect term is dropped from equation (3) and a so-called α^2-dynamo can be obtained, which depends only on the nonlinear α-effect. These dynamos tend to produce steadily growing fields. If the differential rotation is much stronger than the α–effect, then the α–effect term is dropped from equation (3) and an $\alpha\omega$-dynamo can be obtained. These dynamos produce oscillatory waves that propagate at right angles to the shear flow. Their propagation toward the poles or toward the equator depends upon the sign of α and the direction of the velocity shear. If the α–effect and the ω–effect are of similar strength, an $\alpha^2\omega$-dynamo can be obtained. These dynamos also produce oscillatory behavior, but with periods that differ from those for $\alpha\omega$–dynamos depending upon the relative strength of the α–effect.

Kinematic dynamos for the Sun have been constructed from these equations by taking a specified rotation profile, $U_\phi(\theta,r)$, and a functional form for α. Dynamos produced in the 1970s reproduced many of the characteristics of the solar cycle. These were $\alpha\omega$–dynamos, in which the Sun's differential rotation takes a poloidal magnetic field and shears it to produce a stronger toroidal field below the surface. This toroidal field is then lifted and twisted by the α–effect to produce a poloidal field of reversed polarity (Figure 2). The key ingredients in these dynamos were a rotation profile in which the rotation rate increases inward and left-handed helicity in the northern hemisphere. These conditions produce dynamo waves that propagate toward the equator in agreement with Spörer's Law. The problem with these dynamos is the constraints they place on the fluid flows. To produce a dynamo with a 22-year period, the effect produced by the convection must be diminished enormously; otherwise, very short cycles result. In addition, the rotation profiles they use do not agree with the helioseismic profiles.

Nonaxisymmetric flows with helicity result from the effects of rotation on convection. As fluid elements rise and expand, the Coriolis force produces a clockwise rotation in the northern hemisphere giving left-handed helicity. Likewise, as fluid elements sink and contract, a counter-clockwise rotation is produced which also gives left-handed helicity. Right-handed helicity would be produced in the Southern Hemisphere (Figure 3).

Self-consistent magnetohydrodynamic dynamos were produced in the 1980s. These large eddy simulation models start with the equations of motion and the induction equation and calculate numerically both the velocity field and the magnetic field. With these models, the convection itself explicitly produces both the differential rotation for the ω–effect and the helicity for the α–effect. The calculated fields are not consistent with the observations. The rotation profile produced in these models has rotation constant on cylinders. While the α effect has the expected sign, the rotation rate decreases radially inward, contrary to the helioseismic observations, and the dynamo waves propagate toward the poles, contrary to Spörer's Law. These dynamos also had short cycle periods due to the large magnitude of the α–effect.

Problems in Dynamo Theory

A major problem shared by both types of dynamos is the nature of the internal rotation profile as determined by helioseismology. Although the magnetohydrodynamical models produce surface rotation profiles in agreement with observations, the internal profiles disagree. Likewise, the internal profiles assumed to be present in the kinematic models disagree with the observations.

This problem extends beyond dynamo theory itself. Dynamical models for the convection zone produce rotation profiles with surfaces of constant

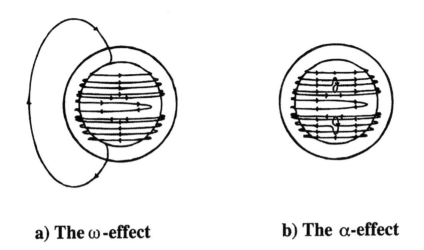

a) The ω-effect   b) The α-effect

Figure 2. The two basic dynamo processes: The ω–effect and the α–effect. With an $\alpha\omega$–dynamo, the ω–effect shown in Figure 2a is produced by differential rotation shearing a north-south field line and wrapping it around the solar interior to produce a strong azimuthal field. The α–effect shown in Figure 2b is produced by helical motions that lift and twist the azimuthal field to produce a new north-south field of opposite polarity.

rotation rate lying on cylinders aligned with the rotation axis. The largest convection eddies become elongated north to south to form banana-shaped cells. Horizontal flows within these cells are turned by the Coriolis force so that eastward momentum is transported toward the equator to maintain the latitudinal differential rotation observed at the surface. While this process is well understood and produces the observed surface profile, the internal rotation profile is all wrong—both for the dynamo and for agreement with the observed internal profile. This remains an outstanding problem in convection zone dynamics.

Another problem shared by both types of dynamos is that magnetic flux tubes should be buoyant and not remain in the convection zone long enough for the fluid motions to work on them. The magnetic pressure within a flux tube requires a smaller contribution from the gas pressure inside to balance the gas pressure outside. A tube in thermal equilibrium with its surroundings gives a lower gas density and makes the tube buoyant.

These two problems have led to the suggestion that the dynamo acts in the interface layer at the base of the convection zone, where flux tubes are less buoyant due to the stable stratification. Helioseismology results show that strong radial shear in the rotation profile occurs in this layer. It is also

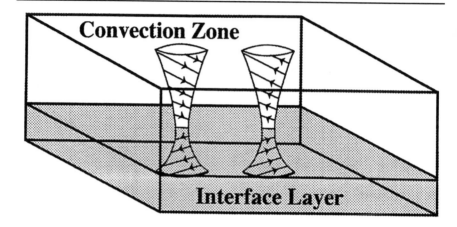

a) Northern Hemisphere

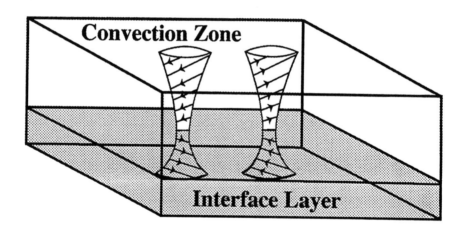

b) Southern Hemisphere

Figure 3. Helicity production by convection in rotating layers. The Coriolis force acting on convective flows produces left-handed helicity in the convection zone in the north (a) and right-handed helicity in the south (b). Converging flows in down-drafts spin counter-clockwise in the north (a) and clockwise in the south (b) while diverging flows in updrafts spin in the opposite directions. The opposite helicity is produced in the interface layer where the flows in downdrafts diverge and flows in updrafts converge.

expected that the more vigorous convective motions will overshoot and penetrate into this layer. Although, for the equatorial region, the rotation rate decreases inward, the α–effect should still have the correct sign. In this interface layer, sinking fluid should expand as it spreads out along the bottom, while rising fluid should contract as the fluid converges in updrafts. This gives right-handed helicity in the northern hemisphere and produces dynamo waves that propagate in accordance with Spörer's Law. In the higher latitudes where the rotation rate increases inward, these waves should move in the opposite direction. Details concerning dynamos in this interface layer have been examined by several investigators. Their models solve some problems associated with the convection zone but produce others.

One of the remaining problems with current models of the solar dynamo is actively being investigated. It concerns magnetic diffusion. For any of these dynamos to work, diffusion is needed so that magnetic fields can reconnect to form new topologies. Ultimately this reconnection must take place in small-scale diffusive processes. The problem is that vigorous small-scale turbulence should amplify the magnetic field to levels that would prohibit the flows from moving the field any further. This limits the amplitude of the mean fields to values less than those observed.

Many of the current efforts in solar dynamo theory are associated with the dynamics of magnetic flux tubes themselves. Several researchers have examined how buoyant flux tubes move through the convection zone. Weak fields tend to rise parallel to the rotation axis and emerge at high latitudes. Fields with strengths of ~100 kG at the base of the convection zone are required to produce sunspots at the observed latitudes. Other investigators are studying the interactions between fluid flows and fibril magnetic field structures. The difficulty of including thin tubes with strong magnetic fields in global models is a severe computational problem for solar dynamo theory.

We still need to know more about the dynamics of the solar convection zone. Helioseismology is our best hope for answering our questions. The Global Oscillations Network Group completed its network of instruments in late 1995, and the European Space Agency and NASA launched the Solar and Heliospheric Observatory with an array of helioseismology instruments in December 1995. These new instruments promise to tell us much more about the solar interior, convection zone dynamics, and the solar dynamo.

References

Foukal, P. V., *Solar Astrophysics*, John Wiley, New York, 1990.

Moffatt, H. K., *Magnetic Field Generation in Electrically Conducting Fluids*, Cambridge University Press, New York, 1978.

Parker, E. N., *Cosmical Magnetic Fields*, Clarendon Press, Oxford, 1979.
Stix, M., *The Sun, An Introduction*, Springer-Verlag, New York, 1989.

D. H. Hathaway
Space Science Laboratory / ES82, NASA Marshall Space Flight Center, Huntsville,
AL 35812

Cosmic Rays

J. R. Jokipii

For many years cosmic rays provided the most important source of energetic particles for studies of subatomic physics. Today, cosmic rays are being studied as a natural phenomenon that can tell us much about both the Earth's environment in space and distant astrophysical processes.

Cosmic rays are naturally occurring energetic particles—mainly ions—with kinetic energies extending from just above thermal energies to more than 10^{20} electron volts (eV). They constantly bombard the Earth from all directions, with more than 10^{18} particles having energies >1 MeV striking the top of the Earth's atmosphere each second. Figure 1 illustrates the continuous cosmic ray energy spectrum.

The broad maximum in the spectrum at 10^8–10^9 eV defines the typical cosmic ray—a proton having approximately 10^9 eV. The abundances of certain unstable isotopes in meteorites show that cosmic rays have been present at nearly their current level for hundreds of millions of years. Studies of electromagnetic waves produced by cosmic rays in distant astrophysical sources reveal that they are present throughout the disk and halo of our galaxy and in other galaxies as well. Cosmic rays are now believed to be produced naturally in astrophysical plasmas, nearly all by a process called diffusive shock acceleration, which occurs naturally at collisionless shock waves. Observations of cosmic rays began in the early part of this century when C.T.R. Wilson, using his cloud chamber, studied the puzzlingly high level of atmospheric ionization. In 1912, with balloon-borne ionization detectors, Victor Hess showed that this ionization increased with increasing altitude and concluded that radiation was coming from above the atmosphere, a discovery which earned him the Nobel Prize in 1936. This myste-

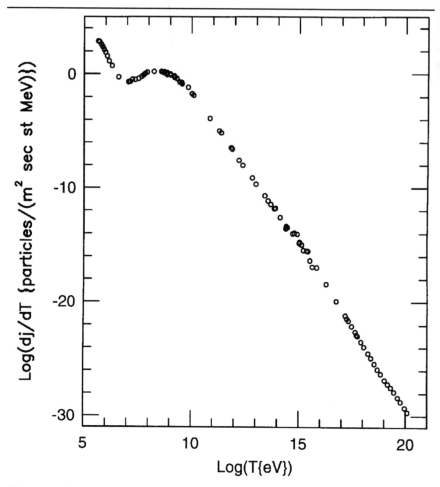

Figure 1. The total observed cosmic-ray intensity as a function of energy. At energies below some 10^{14} eV this is principally hydrogen.

rious radiation was subsequently given the name "cosmic radiation," and it has been the object of intense scientific interest ever since.

Most of the observed cosmic-ray particles originate outside the solar system and are called galactic cosmic rays. The solar wind also produces a variety of energetic particles, the most energetic of which consists of the so-called anomalous cosmic rays [*Mewaldt et al.*, this vol.]. They are believed to be accelerated by the termination of the solar wind. The Sun also sporadically accelerates many particles with a spectrum extending occasionally beyond 10^9 ev [*Lin*, this vol.]. Except for brief periods just after solar flares, or perhaps near strong shock waves, the cosmic rays are observed to be nearly isotropic, meaning that they are distributed nearly uniformly in direction.

Traditionally, cosmic rays are considered to be those particles having energies of the order of a MeV or higher. However, recent observations show that the cosmic-ray or energetic-particle distribution is generally an approximate power law (in momentum) continuation of the thermal distribution. There is no particular energy at which there is a gap between thermal particles and cosmic rays.

Cosmic rays comprise a central phenomenon in solar-terrestrial science and their study is important for many reasons. They provide an increasingly valuable probe of the solar wind and heliosphere, particularly in regions not readily accessible to in situ observations. Cosmic rays are a major source of the trace radioactive components of the atmosphere and extraterrestrial samples used in age dating. Cosmic-ray interactions with the atmosphere may have implications for various aspects of climate and weather. Furthermore, cosmic rays are part of collisionless plasmas throughout the universe, and better understanding of them will help us understand astrophysical plasmas. Finally, cosmic rays can provide information about the origins of matter. If the energy density of cosmic rays is high enough, they can affect the dynamics of the plasmas in many ways.

Galactic Cosmic Rays

Galactic cosmic rays are thought to be accelerated primarily at supernova blast waves, which are vast quasi-spherical shock waves that result from supernova explosions. The typical cosmic ray is then confined to the galaxy by the galactic magnetic field for a few tens of millions of years, enough time to cross the galaxy thousands of times.

The mean confinement time of these "primary" cosmic rays in the galaxy is obtained from the "secondary" cosmic rays, which result from the rare collisions of energetic primary cosmic rays with interstellar gas particles. The number of these secondaries depends on the amount of matter traversed by the primary cosmic rays since their creation. This, in turn, depends on both the age of the cosmic rays and the average density of matter where they propagate. Furthermore, some of these secondary cosmic rays are unstable atomic nuclei that decay as they propagate around the galaxy. In particular, the isotope ^{10}Be has a decay half-life of about a million years. Comparison of how much ^{10}Be is in the flux of cosmic rays with other stable nuclei allows determination of the average age of the cosmic rays. Since the ^{10}Be flux is smaller than expected if it all has survived, much of it has decayed. From this it follows that the mean lifetime is larger than the

[10]Be decay lifetime, or some 10 m.y. Then, one deduces that the average density of gas in the region where the cosmic rays propagate is significantly less than that of the galactic disk, implying that the cosmic rays spend much of their life in the rarefied regions above the disk, called the galactic halo. These cosmic rays contain enough energy that they are a major factor in shaping the galactic disk and halo.

The Heliosphere and Cosmic Rays

The heliosphere is a spheroidal cavity, some 200 AU across, formed in the interstellar gas by the radially flowing solar wind [*Axford and Suess*, this vol.]. We now know that the heliosphere has at least two effects on cosmic rays: It significantly distorts the flux of galactic cosmic rays having energies less than about 10^{12} eV coming in from outside, and the heliosphere itself produces particles with energies up to a few GeV. In the inner heliosphere cosmic rays contribute a negligible amount to the local plasma energy and furthermore, they do not interact collisionally with the ambient gas particles. They are also far too energetic for gravity to have any significant effect. In this case they can be treated as a distinct population of individual energetic particles coexisting with the background plasma. In the outer heliosphere, where the wind is much weaker, cosmic rays may have enough energy to alter the dynamics of the flow.

Structure of the Heliosphere

The solar wind flows radially outward from the Sun at several hundred kilometers per second from about 10 solar radii out to about 100 AU [*Goldstein*, this vol.], where it slows down because of the resistance of the interstellar medium. Here it passes through a spheroidal shock wave called the solar-wind termination shock. The magnetic field is frozen into the moving plasma, so the radial flow drags the solar magnetic field outward, with solar rotation causing it to assume the shape of an Archimedean spiral. The radial convection of this spiral magnetic field creates an electrostatic electric field, which causes an electrostatic potential difference between the heliospheric pole and equator of approximately 250 million volts. Finally, a turbulent spectrum of magnetic irregularities caused by turbulence in the solar wind is superimposed on the Archimedean spiral magnetic field.

The interplanetary magnetic field has a remarkable structure that varies with the sunspot cycle. The field is an Archimedean spiral in one direction (say, outward from the Sun) in the northern hemisphere and in the opposite direction in the southern heliosphere. The two oppositely directed fields are separated by the thin, warped heliospheric current sheet. This sheet is at its flattest—oscillating between ±10° or so above and below the

Sun's equatorial plane—during sunspot minimum, and its oscillations increase in amplitude approaching sunspot maximum. The structure becomes much more complex near sunspot maximum, but approaching the next sunspot minimum, it regains its simple structure, but with the direction of the magnetic field reversed. The field in the northern hemisphere was outward during the 1976 sunspot minimum, and inward during the 1987 minimum. Thus we have an 11-year sunspot cycle and a 22-year solar magnetic cycle.

The structure of the extreme outer heliosphere beyond the termination shock is not well determined, primarily because it has not yet been observed directly. The exploration of this region of space is one of the highest priorities of space plasma physics. We expect that the general picture of cosmic rays in the inner heliosphere is not very sensitive to the details of the wind in its outermost parts.

Solar Modulation of Galactic Cosmic Rays

The solar wind excludes nearly all of the galactic cosmic rays with energies below a few hundred MeV from the inner heliosphere and significantly affects even higher energies, up to about 10^{12} eV. This solar modulation of the galactic cosmic-ray flux mainly reduces the flux of cosmic rays near sunspot minimum. However, it has a variety of other effects related to both the 11-year sunspot cycle and the 22-year solar magnetic cycle [*Hathaway*, this vol.]. Longer-term variations also occur, but these must be inferred indirectly from proxy indicators such as unstable isotopes—for example, carbon 14—generated by cosmic rays striking the Earth's atmosphere.

This solar modulation of the galactic cosmic-ray flux is a consequence of the transport of cosmic rays in the outwardly moving solar wind. The cosmic-ray transport theory used today to describe this phenomenon was first written down by E.N. Parker in 1965 as the combination of four basic physical effects. Because the wind is so rarefied, collisions of the cosmic rays with the ambient solar-wind particles are completely negligible, so the cosmic rays respond only to the convected magnetic field of the solar wind.

The cosmic-ray particle may be thought of as gyrating rapidly about the spiral magnetic field and moving at the same time parallel to it. Magnetic irregularities occasionally "scatter" them randomly in angle, rapidly enough to maintain near-isotropy of the distribution in pitch angle in the coordinate frame moving with the flow. The particles consequently random-walk, or diffuse, relative to the convecting solar wind. The particles experience two additional effects while this is going on. First, the solar-wind plasma is generally either expanding as it flows outward from the Sun or is compressing at shocks. The magnetic irregularities are there-

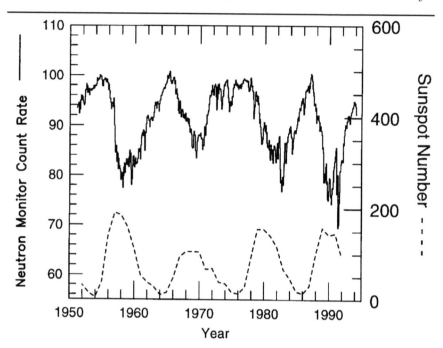

Figure 2. The solid line is the counting rate of the climax neutron monitor over the last four decades. The dotted line is the sunspot number over the same period. (Courtesy of Roger Pyle)

fore moving apart or closer together, so the cosmic rays undergo an "adiabatic" cooling or heating. Second, because the gyromotion around the spiral field is generally faster than the scattering, the particles undergo a drift caused by the large-scale spatial variation of the spiral field. The Parker transport theory combines these four effects and is very widely used in the discussion of energetic particle transport in astrophysical plasma flows.

The reduction of the galactic cosmic-ray intensity is most significant during each sunspot maximum, giving rise to an 11-year variation of the cosmic-ray intensity which is inversely related to solar activity. Figure 2 illustrates the variation in the cosmic-ray intensity over the past few sunspot cycles and, for comparison, the sunspot number over the same period. The 11-year variation is quite apparent. The relative amplitude of this variation decreases with particle energy. Also apparent in Figure 2 is an alternation in the shape of the cosmic-ray maxima from one maximum to the next, from a sharply peaked maximum in 1965 to a flatter maximum in 1976. This latter effect is thought to be one of a number of 22-year, cosmic-ray variations which are probably related to the change in sign of the solar and interplanetary magnetic field.

It now appears that these 22-year and 11-year cosmic-ray variations may be understood as a consequence of two different physical effects. During most of the sunspot cycle, centered on each sunspot minimum (cosmic-ray maximum), the distribution of cosmic rays reflects primarily a balance between the inward drift motions of the particles and the cooling in the expanding solar plasma. This spatial distribution is considerably different for the two opposite directions of the interplanetary magnetic field, leading to a 22-year variation. On the other hand, near each sunspot maximum, the magnetic field is disordered, and the cosmic rays are affected primarily by outward-moving solar disturbances. Large depressions in intensity occur in association with very large disturbances in the solar wind, causing the deepest depressions in the intensity every 11 years, at sunspot maximum.

We may also understand the differences in the shapes of successive cosmic-ray maxima in the framework of the above picture. During the years around alternate sunspot minima (such as in 1987 or 1965), the drift motions bring positively charged cosmic-ray ions into the inner heliosphere via the equatorial heliosphere. In this case, as solar activity begins to build, it starts to change the equatorial parts of the solar wind (most important, the latitude excursion of the current sheet begins to increase). Because the particles at this phase must travel through this disturbed equatorial region, they encounter increasing difficulty in getting in, and the observed intensity begins to decrease soon after solar activity begins to increase. Hence, the cosmic-ray maximum is short and sharply peaked. On the other hand, during the other sunspot minima (1976 and 1998), the particles drift in via the polar regions of the heliosphere and are less sensitive to the increase in solar activity near the equator. This results in a longer (flatter) maximum in the cosmic-ray intensity. Thus the intensity does not drop significantly until solar activity is more global in nature and extends to the polar regions.

It has become apparent over the last decade that the same transport equation that successfully explains many of the properties of the modulation of galactic cosmic rays also predicts that the particles will be efficiently accelerated at shock waves. This acceleration at the termination shock can significantly affect the modulation of ~GeV galactic cosmic rays, but it does not change the general nature of the picture discussed above. However, it does give rise to a completely new species of lower-energy cosmic rays whose existence was not discovered until the early 1970s—the anomalous cosmic rays.

Acceleration of Cosmic Rays

The acceleration of most cosmic rays is now believed to occur primarily at shock waves propagating through astrophysical plasmas. This acceleration occurs because of the extremely rapid compression of the plasma as the shock front passes over it. This compression of the ambient plasma results in an increase in the energy of the fast charged particles, but only by a small fraction in each encounter with the shock. Most particles encounter the shock a small number of times and are accelerated only a small amount, while others are fortunate enough to cross the shock many times, and gain many times their initial energy. So a few particles gain a significant amount of energy. This acceleration process, called diffusive shock acceleration, is extremely efficient and produces an energy distribution which is the same for a very wide range of conditions and which is very close to the energy distribution observed.

This acceleration has been observed directly at the Earth's bow shock and propagating shock waves in the interplanetary gas. It is thought to occur at supernova blast waves, where it produces the majority of galactic cosmic rays. In addition, a newly observed species of cosmic rays, called anomalous cosmic rays, is now thought to be accelerated at the termination shock of the solar wind [*Mewaldt et al.*, this vol.]. Acceleration of charged particles by propagating interplanetary shocks is frequently observed, but because these shocks tend to be significantly weaker and shorter-lived than the termination shock, they produce particles only up to some tens of MeV energy.

Cosmic Electrons

Cosmic electrons, with an intensity of approximately 1% of the intensity of nucleons, are more than merely the electron component of cosmic rays. Because of their low mass, they provide a different kind of information regarding the sources and regions of propagation. In addition, the positron component is produced copiously in the collisions of nucleonic cosmic rays with ambient matter and therefore is also a valuable diagnostic of propagation. Electrons and positrons—again as a consequence of their low mass—emit radio waves as they gyrate in the interstellar magnetic field. This permits the study of cosmic rays in distant sources. Partly because of their low intensity relative to ions, our knowledge of the acceleration and transport of cosmic electrons is still very poor, and there are no generally accepted theories of their acceleration and transport. In principle, diffusive shock acceleration of energetic electrons can operate as well as for ions, although the details are different. The time variation of the electron component during the last cosmic-ray maximum in 1986 showed a counting rate of ~5 GeV electrons—much higher than in the previous cosmic-ray maximum, which cannot be easily fit into current modulation theory. Observations of the cosmic-ray electron component will contribute greatly to our knowledge. Obtaining such observations is of the highest priority.

References

Lin, R. P., Exploring the enigma of solar energetic particles, *Eos, Trans. AGU*, 75, 457, 1994.

Mewaldt, R. A., A. C. Cummings, and E. C. Stone, Anomalous cosmic rays: Interstellar interlopers in the heliosphere and magnetosphere, *Eos, Trans. AGU*, 75, 185, 1994.

J. R. Jokipii
Department of Planetary Science, University of Arizona, Tucson, AZ 85721

Anomalous Cosmic Rays: Interstellar Interlopers in the Heliosphere and Magnetosphere

R. A. Mewaldt, A. C. Cummings, and E. C. Stone

Since the beginning of the space age, it was known that two main populations of energetic particles pervade interplanetary space: Galactic cosmic rays (GCRs) [*Jokipii*, this vol.], which originate in sources outside the solar system, and solar energetic particles (SEPs), associated with transient events on the Sun [*Rust*, this vol.; *Lin*, this vol.].

But over 20 years ago, instruments on the Pioneer 10, IMP 5, and IMP 7 spacecraft discovered a third component of energetic particles known as "anomalous cosmic rays" (ACRs), that represents a sample of nearby interstellar material.

Over the past 2 decades, ACRs have been used to study the acceleration and transport of energetic particles within the solar system, deduce the global properties of the heliosphere [*Axford and Suess*, this vol.]—the bubble of solar wind that envelopes the solar system, and study the interstellar material itself.

It has recently been shown that some of these ACRs become trapped in Earth's magnetic field, where they form a radiation belt composed of interstellar material (Figure 1). Also, ACRs are being used to address a question that has existed ever since the discovery of the solar wind: "How large is the heliosphere?"

Discovery of ACRs

The unusual composition of ACRs led to their discovery in 1973. Pioneer and IMP observations revealed anomalous excesses of several elements in low-energy cosmic rays, including He, N, O, and Ne. For example, O

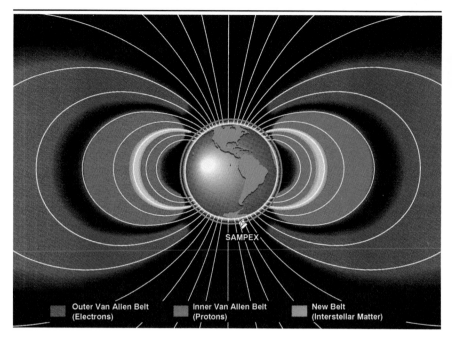

Outer Van Allen Belt   Inner Van Allen Belt   New Belt
(Electrons)   (Protons)   (Interstellar Matter)

Figure 1. Schematic cross section of the trapped radiation belts surrounding the Earth. The Van Allen belts are shown in blue and purple. The inner belt is composed mainly of energetic protons, while the outer belt is mainly energetic electrons. A newly identified radiation belt, shown as two bright yellow crescents, is composed of energetic heavy nuclei that originated in the local interstellar medium. All of these belts approach closest to Earth in the south Atlantic region because of the off-set of the Earth's magnetic dipole. The orbit of the polar-orbiting SAMPEX satellite, which has been studying the new belt, is indicated.

exceeded C in abundance by about 30 to 1, and He was more abundant than H (see Figure 2). By contrast, in SEPs and GCRs, C and O are almost equally abundant, and H is typically ≥10 times more abundant than He.

In 1977, the Pioneer 10 and 11 spacecraft were joined in their journey to the outer solar system by Voyager 1 and 2. As the four spacecraft moved outward, the number of particles hitting 1 square centimeter each second increased, indicating that ACRs were not accelerated on the Sun.

In addition, the ACR intensity was found to be inversely correlated with the 11-year sunspot cycle, similar to the well-known "solar modulation" of GCRs. However, while the low-energy GCR intensity varies by a factor of <10 from solar minimum to solar maximum, depending on particle energy, the intensity of ACRs varies by a factor of >100! Both observations indicated that the source of ACRs must be well beyond the Pioneer and Voyager spacecraft.

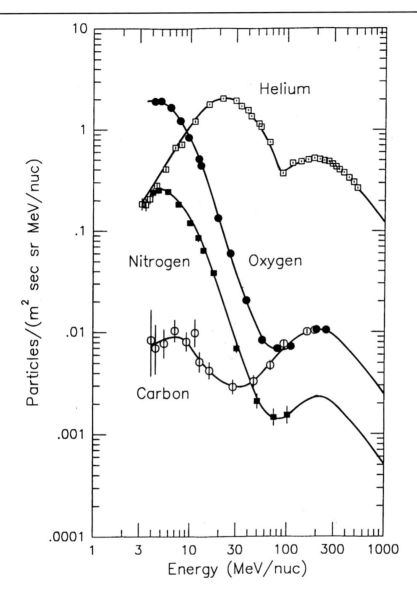

Figure 2. Intensity of cosmic ray He, C, N, and O nuclei as a function of their measured kinetic energy. These data were measured by instruments on Voyager 2 at a distance of ~23 AU from the Sun during the 1987 solar minimum. At kinetic energies >100 MeV/nuc, the particles are of galactic cosmic ray origin. The enhancements in the spectra of He, N, and O below ~50 MeV/nuc are due to anomalous cosmic rays. A somewhat smaller ACR contribution is also observed for C.

The Origin of ACRs

Soon after the discovery of ACRs, *Fisk et al.* [1974] proposed that they represent a sample of particles from interstellar space. To understand this suggestion, the interaction of the interstellar medium (ISM), the vast region between the stars, and the heliosphere must be considered.

As the Pioneer and Voyager spacecraft moved to the outer solar system, they studied the solar wind at increasing distances from the Sun. The solar wind is the extension of the Sun's hot corona. It expands at high velocity into interplanetary space and is composed of protons, electrons, and heavier ions moving radially outward at supersonic speeds of ~400 km/s [*Goldstein*, this vol.].

Embedded in the solar wind is the distended solar magnetic field. As the solar wind expands, it blows a bubble, pushing against the interstellar magnetic field and the thin gas of the local ISM.

Well beyond the orbit of Pluto is an interface called the heliopause, which separates the bubble of solar wind plasma from the ISM (Figure 3). As the solar wind approaches the heliopause, it slows abruptly to subsonic speeds, forming the solar wind termination shock.

As the Sun moves through interstellar space, the heliosphere encounters the gas that makes up the ISM. Interstellar ions, having lost one or more of their electrons, are prevented from flowing across the heliopause by the heliospheric magnetic field. However, the electrically neutral atoms are unaffected by the magnetic field and can drift into the inner heliosphere, where some of the atoms will be ionized by solar UV radiation or by charge exchange with the solar wind (Figure 4).

These new ions are then picked up by the solar wind and convected into the outer heliosphere. *Fisk et al.* [1974] suggested that these "pickup ions" are accelerated to velocities 10–20% of the velocity of light in the outer heliosphere to become ACRs.

Pesses et al. [1981] proposed that this acceleration takes place at the solar wind termination shock [*Jokipii*, this vol.]. Once accelerated, some of these particles diffuse and drift into the inner heliosphere as cosmic rays subject to the same solar cycle modulation as GCRs.

The model described explains the unusual composition of ACRs, because it selects only those elements that are predominantly neutral in the ISM. Atoms that are difficult to ionize, with first ionization potential (FIP) ≥ 13.6 eV (e.g., He, N, O, Ne, and Ar) are primarily neutral in the ISM, while those with FIP <13.6 eV (e.g., C, Mg, Si, and Fe) are expected to be ionized to an overwhelming extent.

All of the common high-FIP elements, including Ar and H, have now been found in ACRs [e.g., *Cummings and Stone*, 1987]. ACR carbon has also been detected, but only in very small amounts, as expected if only ~1% of interstellar C is in the neutral state. These measurements have been used to obtain

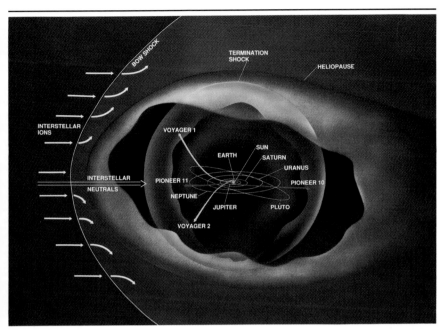

Figure 3. Schematic illustration of the large-scale structure of the heliosphere. Surrounding the solar wind termination shock is the heliopause, which forms the interface between solar and interstellar plasma, and a possible "bow shock" that may be located beyond the heliopause. The present positions of the Pioneer and Voyager spacecraft are indicated. The termination shock is assumed to be 67 AU from the Sun, as derived from an analysis of 1987 anomalous cosmic ray data from Pioneer 10 and Voyager 2.

the relative abundances of neutral atoms in the local ISM.

The model in Figure 4 predicts that ACRs should be singly charged, having lost only a single electron. By contrast, GCRs have essentially been stripped of electrons during their ~10-million-year passage through the galaxy.

Several aspects of this picture have been verified. The existence of neutral H and He flowing into the heliosphere was established in the early 1970's. Recently, the Ulysses spacecraft directly measured neutral He atoms streaming into the inner heliosphere. In addition, the Active Magnetospheric Particle Tracer Explorer and Ulysses missions have both observed "pickup" ions with the expected properties [e.g., *Geiss et al.*, 1994].

Exploring the Heliosphere with Anomalous Cosmic Rays

The fleet of Pioneer and Voyager spacecraft provided our first exploratory view of the three-dimensional structure of the heliosphere. One of their

Origin of Anomalous Cosmic Rays

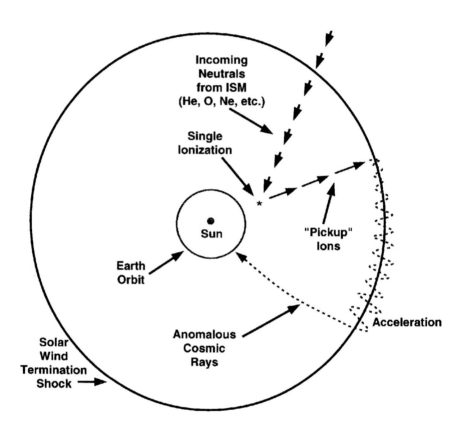

Figure 4. The origin of anomalous cosmic rays.

key objectives is to locate the termination shock and heliopause, thereby defining the size of the heliosphere.

An estimate of the distance to the termination shock has been obtained by extrapolating the ACR intensity gradients measured between Voyager 2 at ~20 AU and Pioneer 10 at ~40 AU into the outer heliosphere [*Cummings et al.*, 1993a]. It was found that the time dependence of the ACR fluxes of He and O during the 1987 solar minimum period could best be explained if the termination shock was at 67±5 AU. The shock is expected to move in response to solar wind conditions and a recent estimate places it at ~85 AU in 1994.

It is interesting to compare this result with estimates of the distance to the heliopause, based on data from the Voyager plasma wave experiments [*Gurnett et al.*, 1993]. Analysis of radio emissions observed by these experiments, coming from the direction of the nose of the heliopause, suggests a heliopause distance of 116–177 AU in 1992 near solar maximum.

Although these shock and heliopause estimates are for different phases of the solar cycle, they appear to be consistent with models of the heliosphere if the heliopause distance is toward the lower end of this range. These findings suggest that Voyager 1 may pass through the termination shock by 2000, when it will be beyond 76 AU. If so, Voyager 1 will be able to observe, in situ, the acceleration of particles to cosmic ray energies.

Trapped ACRs

The well known Van Allen radiation belts [*van Allen*, this vol.] are regions of intense fluxes of energetic protons and electrons trapped in Earth's magnetic field. *Blake and Friesen* [1977] predicted that if ACRs are indeed singly charged, they could also become trapped in the magnetosphere, forming a radiation belt of trapped ACRs.

Because a singly charged oxygen ACR has a radius of gyration eight times larger than fully ionized oxygen as it spirals in the Earth's magnetic field, it can penetrate to lower latitudes than if it were fully stripped. If it were then to skim the atmosphere, the ACR oxygen ion would be stripped of all its electrons, reducing its radius of gyration and leaving it to spiral back and forth from one hemisphere to another.

The first evidence for the existence of trapped ACRs was provided by a team of Russian and U.S. scientists using observations from a series of COSMOS satellites launched from 1985 to 1988 [*Grigorov et al.*, 1991]. The detectors on COSMOS could not directly measure the location of the trapped O because there was no record of where or when the particles were observed during the spacecraft orbit. However, they did reveal that the intensity of trapped O tracked the interplanetary intensity of ACR nuclei measured by IMP 8, increasing from 1986 to early 1987, and then decreasing dramatically in intensity with the onset of solar activity in mid-1987.

This study also confirmed that ACR nuclei originated as "pickup" ions. COSMOS measurements of cosmic rays able to penetrate through Earth's magnetic field were compared with simultaneous measurements in interplanetary space from IMP 8 and ICE. It was found that ACR oxygen nuclei must have had access to lower latitudes than either SEP or GCR oxygen of the same energy, consistent with the prediction that ACR nuclei had lost only a single electron [*Adams et al.*, 1991].

SAMPEX Mission to Study ACRs in Depth

In July 1992, SAMPEX was launched into a polar orbit carrying instruments designed to study ACRs in the Earth's environment. During its first year, SAMPEX confirmed that ACRs are singly charged, and it has located a narrow belt of trapped ACRs within the inner of the Van Allen radiation belts [*Cummings et al.,* 1993b, see Figure 1].

SAMPEX found the third radiation belt to be composed of oxygen with smaller amounts of N and Ne and very little C. The observed relative abundances of C to N to O are consistent with those of ACRs. The striking near-absence of C in this third radiation belt is inconsistent with other possible sources of trapped particles such as the solar wind, SEPs, or GCRs.

Besides confirming Blake and Friesen's prediction and the COSMOS observations, the SAMPEX data show that the location of the belt is much closer to the Earth than expected and that it is much narrower.

In essence, the new radiation belt corresponds to a magnetic bottle that holds a sample of interstellar material. However, because the bottle has a leak—the ions slowly lose energy to the thin upper atmosphere—the material in the bottle is a balance between this leak and the rate at which it is being filled. The intensity of ACR oxygen inside the bottle is about 100 times greater than in interplanetary space.

The rate of flow into the bottle varies with the interplanetary ACR intensity, and as a result, the intensity of ions trapped in the bottle varies by perhaps a factor of ~1000 over the solar cycle.

SAMPEX observations over the next few years should provide an opportunity to use this third radiation belt to study magnetospheric processes and to examine the elemental and isotopic composition of a sample of interstellar matter that, on a Galactic scale, is located right in our own back yard.

References

Adams, J. H., Jr., et al., The charge state of the anomalous component of cosmic rays, *Astrophys. J. Lett., 375,* L45, 1991.

Blake, J. B., and L. M. Friesen, A technique to determine the charge state of the anomalous low-energy cosmic rays, *Proc. 15th Int. Cosmic Ray Conf. (Plovdiv), 2,* 341, 1977.

Cummings, A. C., and E. C. Stone, Composition, gradients, and temporal variations of the anomalous cosmic ray component, in Proceedings of the 6th International Solar Wind Conference, edited by V. J. Pizzo, T. E. Holzer, and D. G. Sime, *NCAR Tech. Note NCAR/TN-306, 2,* 599, 1987.

Cummings, A. C., E. C. Stone, and W. R. Webber, Estimate of the distance to the solar wind termination shock from gradients of anomalous cosmic ray oxygen, *J. Geophys. Res., 98,* 15,165, 1993a.

Cummings, J. R., A. C. Cummings, R. A. Mewaldt, R. S. Selesnick, E. C. Stone, and T. T. von Rosenvinge, New evidence for geomagnetically trapped anomalous cosmic rays, *Geophys. Res. Lett., 20,* 2003, 1993b.

Fisk, L. A., B. Kozlovsky, and R. Ramaty, An interpretation of the observed oxygen and nitrogen enhancements in low-energy cosmic rays, *Astrophys. J. Lett., 190,* L35, 1974.

Geiss, J., G. Gloeckler, R. von Steiger, A. B. Galvin, and K. W. Ogilvie, "Interstellar Oxygen, Nitrogen, and Neon in the heliosphere, *Astronomy and Astrophysics, 282,* 924, 1994.

Grigorov, N. L., M. A. Kondratyeva, M. I. Panasyuk, Ch. A. D. Tretyakova, J. H. Adams, Jr., J. B. Blake, M. Shulze, R. A. Mewaldt, and A. J. Tylka, Evidence for trapped anomalous cosmic ray oxygen ions in the inner magnetosphere, *Geophys. Res. Lett., 18,* 1959, 1991.

Gurnett, D. A., W. S. Kurth, S. C. Allendorf, and R. L. Poynter, Radio emission from the heliopause triggered by an interplanetary shock, *Science, 262,* 199, 1993.

Pesses, M. E., J. R. Jokipii, and D. Eichler, Cosmic ray drift, shock wave acceleration, and the anomalous component of cosmic rays, *Astrophys. J. Lett., 246,* L85, 1981.

R. A. Mewaldt, A. C. Cummings, and E. C. Stone
Downs Laboratory of Physics, California Institute of Technology, Pasadena, CA 91125

The Outer Heliosphere

W. I. Axford and S. T. Suess

In explaining and describing the forces that shape the bubble of solar wind surrounding the Sun, there is a dearth of information. But observations from space are alleviating this situation. Three spacecraft moving away from the Sun—Pioneer 10 and Voyagers 1 and 2—are expected to penetrate the boundaries of the heliosphere within the next few years. All three spacecraft first passed close to Jupiter, and now their extended missions have become explorations of the outer heliosphere.

The boundaries of the heliosphere are a standing "termination shock" in the solar wind surrounding the Sun and the "heliopause," dividing the solar wind from the local interstellar medium. Uncertainties about the size and shape of these boundaries make it difficult to estimate exactly the time when the spacecraft will pass them. The termination shock may be nearly spherical or highly elongated, depending on how fast the local interstellar medium is flowing past the heliosphere. Pioneer 10, traveling downstream from the oncoming interstellar wind, may reach the termination shock first if, in fact, the shock is spherical. If the shock is elongated, having a larger dimension in the downstream direction, then Voyagers 1 and 2, traveling upstream, will encounter the shock first. Once these two spacecraft reach the termination shock, they will then pass through a region of solar wind plasma that has been heated by the shock. After a few years, they will pass the heliopause and go into the interstellar medium.

The Heliosphere

In 1955, Leverett Davis first suggested the existence of the heliosphere and its boundaries. He postulated that "solar corpuscular radiation," later named the "solar wind" by Eugene Parker, would force matter and magnetic flux in the local interstellar medium outward, thereby partially excluding cosmic rays. The simplest expression of the concept is that the solar wind blows a spherical bubble that continually expands over the lifetime of the solar system. The expansion only stops if it meets a significant pressure in the interstellar medium. The resulting bubble is characterized by a radius, R, at which the pressure of the solar wind, dominated by its dynamic pressure, is equal to the total interstellar pressure. Using the scant information available 40 years ago, Davis estimated that R was from 200 to 2000 AU.

This simple view did not answer several questions. If the heliosphere is in a steady state, what happens to the accumulated solar wind plasma and magnetic field? How does the supersonic solar wind "feel" the presence of the interstellar medium? What happens if the interstellar plasma is moving with respect to the Sun? The resolution of these problems led to our present understanding of the heliosphere.

Conservation of mass and pressure require that the solar wind be slowed as it meets the interstellar medium. The adjustment occurs at a standing shock wave moving toward the Sun at the same speed at which the solar wind flows away from the Sun. Thus the standing shock wave remains at a fixed distance, R, from the Sun. Using values from Table 1 for the solar wind speed, ion density (which decreases as the inverse square of distance from the Sun), and ion mass, R is calculated to be approximately 85–100 AU.

An analogy for the termination shock can be made with fluid flow in the presence of gravity. If one allows a steady stream of water from a faucet to impinge on a horizontal dinner plate, the water flows radially from the point of impact at "supersonic" speed. Provided the flow rate is not too large, a hydraulic jump or shock wave is produced at a position such that the water outside the shock is just deep enough to flow over the raised edge of the plate. The cartoon in Figure 1 shows that the flow is "subsonic" in this region. In this analogy, the depth of the water can be compared with either the pressure or density of solar wind plasma.

The heliosphere is probably not at rest, however, and any motion of the Sun and the heliosphere in relation to the very local interstellar medium (VLISM) has a dramatic effect on the termination shock and the heliosphere shape. In Figure 2, which illustrates the streamlines of a simple model incorporating VLISM motion, interstellar flow turns at the stagnation point to go around the heliosphere, and the subsonic solar wind flow is turned in the "heliosheath" between the termination shock and the heliopause to flow down the "heliotail." The solar wind inside the termination shock is flowing supersonically and radially. The terminology is closely analogous to that of the Earth's magnetosphere because the two configurations are similar.

Table 1. Properties of the Very Local Interstellar Medium

Property	Value
Neutral Component	
Flow speed	25 ± 2 km/s
Flow direction	75.4° ecliptic longitude
	–7.5° ecliptic latitude
Hydrogen density	0.10 ± 0.01 cm^{-3}
Helium density	0.10 ± 0.003 cm^{-3}
Hydrogen temperature	$(7\pm2)\times10^3$ K
Helium temperature	$(7\pm2)\times10^3$ K
Ionized Component	
Electron density	<0.3 cm^{-3}
Flow speed component	Assumed same as neutral
Flow direction component	Assumed same as neutral
Ion temperature component	Assumed same as neutral
Magnetic Field	
Magnitude	0.1–0.5 nT
Direction	Unknown
Cosmic Rays	
Total pressure	$(1.3\pm02)\times10^{-12}$ dynes cm^{-2}

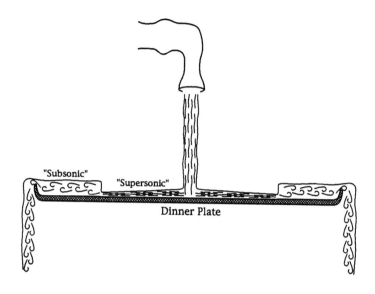

Figure 1. Analog to solar wind flowing into the interstellar medium using water flowing over a dinner plate. The termination shock is represented by the jump between "supersonic" and "subsonic" flow.

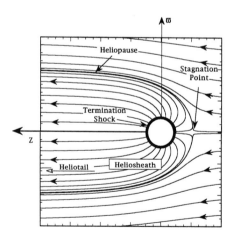

Figure 2. The heliospheric configuration resulting from the interaction between a subsonic interstellar wind and the outflowing supersonic solar wind. The z axis is along the heliotail, opposite to the inflowing interstellar wind and the flow is axisymmetric about this axis.

The Interstellar Medium

Until 1972, most techniques for observing the VLISM required averaging measurements over distances of 100 parsecs or more (a parsec is 3.1×10^{13} km, or 207,000 AU). The average electron density (~0.03 cm^{-3} and magnetic field strength (~0.3 nT, with large fluctuations) were deduced from the observations of Faraday rotation and dispersion of pulsar emissions. Conversely, radio observations of the 21-cm interstellar line had suggested the density of neutral hydrogen to be approximately 0.7 cm^{-3}, a considerably higher value than the 0.1 cm^{-3} value obtained from measurements of the absorption by intervening interstellar matter of Lyman alpha radiation emitted by nearby early type stars.

Although we do not yet have improved values for the electron density and magnetic field strength, the situation with regard to neutral hydrogen and helium has greatly improved since 1972. Instruments on the spacecraft OGO 5 measured the backscattering of solar Lyman alpha radiation by neutral hydrogen atoms as they entered the heliosphere. The backscattered radiation was observed to come from within 5 AU of the Sun and from a direction close to the ecliptic plane. This observation is surprising because the Sun does not move in this direction with respect to nearby stars. The heliocentric distance of maximum backscatter, and hence penetration of the neutral atoms, is determined easily by measuring the parallax of the backscattering from the vicinity of the Earth at different times during the year.

The forces acting on the neutral hydrogen approaching the Sun are gravity and radiation pressure. The atoms are lost when they exchange electrons with solar wind protons or when they are photoionized. The resulting new protons are taken up by the solar wind—hence they become "pickup ions" [*Mewaldt et al.*]. Interstellar atoms traveling at a relative speed of 25 km/s can penetrate the heliosphere approximately 4 AU from the Sun, before being lost by charge exchange.

This distance corresponds roughly to the distance obtained from the OGO 5 parallax measurements. The scattered solar Lyman alpha distribution from a given interstellar neutral hydrogen distribution can be computed and compared with the OGO observations, for example. However, the inflow speed, U, of the neutral atoms, and the loss mean free path, λ, are difficult to determine separately. Hence, the value deduced for U is uncertain due to the difficulties in evaluating λ.

The case of interstellar neutral helium is different because radiation pressure has a negligible effect, and the value of λ for helium is such that the neutral beam penetrates to well within 1 AU if U is 25 km/s. The interesting conclusion is that the maximum intensity of scattered solar helium 584 Å radiation occurs in the downwind direction relative to the motion of the incoming interstellar matter (see Figure 2)—not in the upwind direction as in the case of hydrogen. This result provides a straightforward means of determining the direction of the interstellar wind. The early results from observations of scattered Lyman alpha and helium 584 Å have been confirmed and refined by the neutral helium experiment on Ulysses since its launch in 1989, and by direct observation of pickup ions on the Active Magnetospheric Particle Tracer Explorer (AMPTE) mission. These observations have yielded accurate values for the speed, direction, and temperature of the helium. A summary of the results from these, and other sources is given in Table 1.

With regard to the cosmic ray pressure in Table 1, the high-energy component enters the heliosphere essentially unimpeded. A low-energy component (<100 MeV), which is recent and not necessarily in equilibrium with locally produced cosmic rays, may arise from a nearby bubble of hot gas in the VLISM. The low-energy component of the cosmic rays would contribute to the total pressure, resulting in a thickened termination shock.

Interstellar pickup ions, including hydrogen, helium, and heavier elements, have an interesting fate. After neutral atoms enter the heliosphere, undergo ionization and are swept away by the solar wind, they are carried out to the termination shock. The small percentage that are accelerated to cosmic ray energies propagate back into the inner heliosphere, where they are observed as "anomalous cosmic rays." This process has recently been confirmed by the observation from AMPTE and Ulysses that anomalous cosmic rays are, indeed, singly ionized.

Soft x-ray observations suggest that the Sun is located within an irregular region of hot (10^6 K), low density (5×10^{-2}cm^{-3}) plasma with dimensions on the order of tens to hundreds of parsecs. This "local bubble" of hot plasma was probably produced by a combination of supernovae and stellar winds associated with a group of O and B stars. However, the VLISM does not have the same characteristics as the larger local bubble since the VLISM contains the neutral atoms—observed through the backscattered solar radiation—which have a temperature of only 7000 K. Thus the Sun is usually considered to be immersed in a region that is relatively cool and dense.

As a balance between the pressures of the VLISM and the local bubble must exist, the VLISM should contain unobserved components of plasma and magnetic field to compensate for the low pressure of the observed neutral atoms. If the pressure were mainly magnetic, it would require B = 0.6 nT, whereas plasma alone would require a density greater than 1 cm^{-3} to produce such a pressure, and the consequent ram pressure of an inflow speed of 25 km/s would put the solar wind termination shock at $R = 30$ AU. Since Pioneer 10 and Voyagers 1 and 2 are now near or beyond 50 AU, this cannot be the case.

The relative abundances of hydrogen and helium atoms in the VLISM in Table 1 suggest that the density of the ionized component (mainly protons) is actually not greater than about 0.3 cm^{-3}. If the termination shock is assumed to lie at 100 AU, the density can be as low as 0.1 cm^{-3} if there is also a magnetic field of 3–5 nT. These conditions can provide a match to the hydrostatic pressure in the local bubble if the latter has been slightly over-estimated by about a factor of 2.

Present Understanding of the Heliosphere

Although one might question whether the plasma in the VLISM is moving with the interstellar neutral atoms—at the same speed of 25 km/s—towards the Sun, the plasma and neutral atoms are assumed to be moving together because the collision time between the neutral atoms and ions is about 1000 years, and the mean free path is 10^3–10^4 AU—a small range relative to any plausible interstellar transit time. Therefore the heliosphere is probably experiencing a flow of interstellar wind past itself, rather than just blowing a huge bubble in the VLISM. This kind of stagnation point flow, described in Figure 2, is now accepted as the most likely configuration for the heliosphere.

The configuration may be slightly more complex, however. The fast-mode speed, cf, (see Table 1) in the local fluff is between 10 and 40 km/s—whereas the speed of sound is 10 km/s. An interstellar wind speed, U, greater than cf requires modification to the configuration shown in Figure 2 in three ways. First, a bow shock would form, much like the bow shock in front of the Earth's magnetosphere [*Cowley*, this vol.]. Second, the gas pressure on the downstream side and flanks of the termination shock would be reduced relative to the pressure on the upstream side, causing the shock to be located farther from the Sun in those directions. Figure 3 shows an example of a calculation from a numerical model which incorporates and confirms these two effects. Third, since the interstellar magnetic field probably would not be aligned with the heliotail, the configuration would no longer be symmetric about the heliotail axis. However, effects of an arbitrary magnetic field are not yet incorporated into the numerical simulations.

The numerical model results in Figure 3 employ parameters like those in Table 1, neglecting the effects of the interstellar magnetic field. Since the distance to the shock in the downstream direction is calculated to be more than twice that in the upstream direction, a spacecraft traveling upstream would meet the shock much sooner than in any other direction. This difference is relevant because, as described, Voyagers 1 and 2 are both traveling in the upstream direction while Pioneer 10 is traveling downstream. The importance of this situation is now obvious from Figure 3. Voyagers 1 and 2, which are

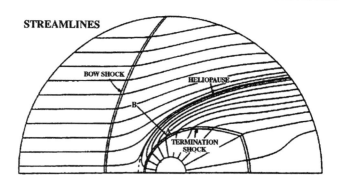

Figure 3. The heliospheric configuration resulting from the interaction between a supersonic interstellar wind and the outflowing supersonic solar wind.

both now more than 50 AU from the Sun, will survive to encounter the upstream shock if it lies inside 120 AU. The heliosheath in the upstream direction—between the termination shock and the heliopause—can be very thin. Voyagers 1 and 2 may therefore pass through the sheath within a few more years and continue into the interstellar medium—an opportunity worth hoping for.

If the interstellar magnetic field is large enough so that $U < cf$, the shock and heliopause and termination shock will more likely have the configuration shown in Figure 2. In this case, Pioneer 10, which is farther from the Sun than Voyagers 1 and 2—but in the downstream direction and moving more slowly than those spacecraft—may be the first to reach the shock. Its relative time of arrival depends completely on whether the shock is spherical, as in Figure 2, or elongated, as in Figure 3.

Because the solar wind pressure depends on both time and location, the termination shock might move inward and outward over the solar cycle by many AU, and by a few AU within the period of a month. Thus the shock may move back and forth over a spacecraft several times after the first encounter.

Termination Shock Distance

Calculations for distances from the Sun to the termination shock and to the heliopause in the upstream direction for a range of interstellar field strengths and densities are shown in Figure 4. The estimated values for R range between 60 and 150 AU, with the values in Table 1, which assume an ionization equilibrium in the VLISM. Because the local bubble is recent, and it would take on the order of 10^6 years to reach equilibrium, this assumption is not necessarily reliable but is used for lack of better information.

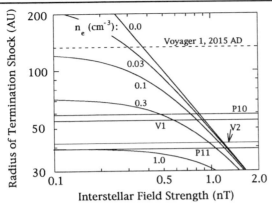

Figure 4. Dependence of upstream distance to the heliospheric termination shock on interstellar magnetic field strength, for five different values for the interstellar plasma density.

If Pioneer 10 or Voyagers 1 and 2 should pass the shock and the heliopause and move into the interstellar medium, the resulting observations would not only add greatly to our knowledge of the heliosphere, they would also supply information on the ionization state in the VLISM. Furthermore, future opportunities for obtaining *in situ* measurements of the interstellar medium are unlikely. A finite window remains, because Pioneer 10 has nearly reached the maximum lifetime of its power supply, and the Voyagers will run out of electrical power by the year 2030.

Several tentative remote detections of the termination shock and heliopause have provided clues to the dimensions of *R*. Probably the most tantalizing evidence is the detection of radio emissions coming from the outer heliosphere. The emissions, thought to come from the nose of the heliopause—the stagnation point in Figure 2—suggest that the distance to the stagnation point is approximately 116–177 AU. These distances place *R* between 60 and 160 AU, an approachable position for the Voyager spacecraft.

The heliosphere is probably well represented by a configuration, similar to that of the Earth's magnetosphere, that is formed by stagnation point flow occurring between the solar wind and an interstellar wind. Inside the heliopause, the solar wind passes through a termination shock that is elongated in the downstream direction and moves back and forth at speeds as high as 100 km/s. The presence of three spacecraft in the outer heliosphere is a unique opportunity to explore the structure and properties of the heliosphere and the VLISM. The indirect evidence indicating that *R* is probably ~100, and possibly as small as 60–70 AU, suggests that the chances of the spacecraft realizing this opportunity are high.

Bibliography

Axford, W. I., The interaction of the solar wind with the interstellar medium, *Solar Wind*, NASA SP308, edited by C. P. Sonett, P. J. Coleman, Jr., and J. M. Wilcox, pp. 609–657, NASA, Washington, D.C., 1972.

Baranov, V. B., Gas dynamics of the solar wind interaction with the interstellar medium, *Space Sci. Rev.*, 52, 90, 1990.

Grzedzielski, S., and D. E. Page (Eds.), *Physics of the Outer Heliosphere, COSPAR Colloquium Series V. 1*, Permagon, New York, 1990.

Holzer, T. E., Interaction between the solar wind and the interstellar medium, *Annu. Rev. Astron. Astrophys.*, 27, 199, 1989.

Parker, E. N., *Interplanetary Dynamical Processes*, Interscience, New York, 1963.

Suess, S. T., The heliopause, *Rev. Geophys.*, 28, 97, 1990.

W. I. Axford
Max Planck Institut für Aeronomie, Postfach 20, Katlenburg Lindau, D 3411 Germany

S. T. Suess
Space Sciences Laboratory/ES82, NASA Marshall Space Flight Center, Huntsville, AL 35812

GLOSSARY

Adiabatic Invariant: In a nearly collisionless, ionized gas, electrically charged particles orbit around magnetic lines of force. Certain physical quantities are approximately constant for slow (adiabatic) changes of the magnetic field in time or in space and these quantities are called *adiabatic invariants*. For example, the magnetic moment of a charged particle, $\mu = mV_\perp^2/(2B)$, is such a constant where $V_\perp$ is the velocity of the particle perpendicular to the magnetic field, B is the magnetic field strength, and m is the particle mass. In a converging field such as in approaching the pole of a dipole magnetic, the field strength increases and therefore $V_\perp$ increases as well because μ has to remain constant.

Aeronomy: The science of the (upper) regions of atmospheres, those regions where dissociation of molecules and ionization are present.

Alfvén Wave (magnetohydrodynamic shear wave): A transverse wave in magnetized plasma characterized by a change of direction of the magnetic field with no change in either the intensity of the field or the plasma density.

Alpha Particle: The nucleus of a helium atom. A positively charged particle, consisting of two protons and two neutrons.

Anisotropic Plasma: A Plasma whose properties vary with direction relative to the ambient magnetic field direction. This can be due, for example, to the presence of a magnetic or electric field. See also Isotropic Plasma; Plasma.

Anomalous Cosmic Ray (ACR): Cosmic ray ions which are singly ionized and which have a distribution in energy different than that for cosmic rays which originate at the Sun or in the milky way outside the solar system. See also Cosmic Ray.

Arcade (coronal, of solar loops): A system of small, arched features connecting bright, compact plages of opposite magnetic polarity. An arcade is a result of emerging bipolar magnetic flux and possibly rapid or continued growth in an active region.

Archimedean Spiral: A plane curve generated by a point moving away from a fixed point while the radius vector from the fixed point rotates at a constant angular rate (i.e. a needle following the groove in a phonograph record).

Astronomical Unit (AU): The mean radius of the Earth's orbit, 1.496×10^{13} cm.

Aurora: A visual phenomenon that occurs mainly in the high-latitude night sky. Auroras occur within a band of latitudes known as the auroral oval, the location of which is dependent on the intensity of geomagnetic activity. Auroras are a result of collisions between atmospheric gases and charged particles (mostly electrons) precipitating from the outer parts of the magnetosphere and guided by the geomagnetic field. Each gas (oxygen and nitrogen molecules and atoms) emits its own particular color when bombarded by the precipitating particles. Since the atmospheric composition varies with altitude, and the faster precipitating particles penetrate deeper into the atmosphere, certain auroral colors originate preferentially from certain heights in the sky. The auroral altitude range is 80 to 500 km, but typical auroras occur 90 to 250 km above the ground. The color of the typical aurora is yellow-green, from a specific transition line of atomic oxygen. Auroral light from lower levels in the atmosphere is dominated by blue and red bands from molecular nitrogen and molecular oxygen. Above 250 km, auroral light is characterized by a red spectral line of atomic oxygen. To an observer on the ground, the combined light of these three fluctuating, primary colors produces an extraordinary visual display. Auroras in the Northern Hemisphere are called the aurora borealis or "northern lights." Auroras in the Southern Hemisphere are called aurora australis. The patterns and forms of the aurora include quiescent "arcs," rapidly moving "rays" and "curtains," "patches," and "veils."

Auroral Electrojet (AE): See Electrojet

Auroral Oval: An elliptical band around each geomagnetic pole ranging from about 75 degrees magnetic latitude at local noon to about 67 degrees magnetic latitude at midnight under average conditions. It is the locus of those locations of the maximum occurrence of auroras, and widens to both higher and lower latitudes during the expansion phase of a magnetic substorm.

Beam (ion, electron, proton): A condition where all of the charged particles move together with the same velocity vector (the opposite to an isotropic distribution).

Beta (e.g., low-beta plasma): The ratio of the thermal pressure to the magnetic 'pressure' in a plasma - $p/(B^2/(8\pi))$ in centimeter-gram-second (c.g.s.) units.

Bi-directional Streaming: Beaming in opposite directions (usually along the magnetic field direction and in the anti-direction).

Blue Jets: Blue jets are narrow geyser-like fountains of blue light that erupt from the tops of some electrically very active thunderstorms and propagate upward to altitudes of 40-50 km at speeds of about 100 km per second. Blue jets seem to occur less frequently than red sprites, but their much longer lifetime of several tenths of a second makes them easier to detect visually when they do occur. Visual sightings of blue jets have been reported by airline pilots. See also Red Sprites.

Bow Shock (Earth, heliosphere): A collisionless shock wave in front of the magnetosphere arising from the interaction of the supersonic solar wind with the Earth's magnetic field. An analogous shock is the heliospheric bow shock which exists in front of the heliosphere and is due to the interaction of the interstellar wind with the solar wind and the interplanetary magnetic field.

Butterfly Diagram: A plot of observed solar active region latitude location vs. time. This diagram, which resembles a butterfly, shows that the average latitude of active region formation drifts from high ($30°$) to low ($10°$) latitudes during a sunspot cycle.

Charge Exchange: An interaction between a charged particle and a neutral atom wherein the charged particle becomes neutral and the neutral particle becomes charged through the exchange of an electron.

Chromosphere: The layer of the solar atmosphere above the photosphere and beneath the transition region and the corona. The chromosphere is the source of the strongest lines in the solar spectrum, including the Balmer alpha line of hydrogen, and the H and K lines of calcium. This is the source of the red (chromium) color often seen around the rim of the moon at the beginning and end total solar eclipses.

Cloud (magnetic): see Magnetic Cloud

Collisional (de-) **Excitation:** Excitation of an atom or molecule to a higher energy state due to a collision with another atom, molecule, or ion. The higher energy state generally refers to electrons in higher energy orbits around atoms.

Convection (magnetospheric, plasma, thermal): The bulk transport of plasma (or gas) from one place to another, in response to mechanical forces (for example, viscous interaction with the solar wind) or electromagnetic forces. Thermal convection, due to heating from below and the gravitational field, is what drives convection inside the Sun. Magnetospheric convection is driven by the dragging of the Earth's magnetic field and plasma together by the solar wind when the geomagnetic field becomes attached to the magnetic field in the solar wind.

Convection Zone: The Sun's convection zone is approximately the outer 30% of the Sun, in which thermal convection is driven by heating from below. The heat comes from the nuclear reactions in the core of the Sun.

Cooling (radiative, thermal): A hot gas or plasma has several ways to cool. One is through ordinary thermal conduction. Another is through radiation of electromagnetic waves.

Coriolis Force: In the frame of a rotating body (such as the Earth), a force due to the bodily rotation. All bodies that are not acted upon by some force have the tendency to remain in a state of rest or of uniform rectilinear motion (Newton's First Law) so that this force is called a "fictitious" forces. It is a consequence of the continuous acceleration which must be applied to keep a body at rest in a rotating frame of reference.

Corona: The outermost layer of the solar atmosphere, characterized by low densities (<10^9 cm^{-3} or 10^{15} m^{-3} and high temperatures (>10^6 K).

Coronal Hole: An extended region of the solar corona characterized by exceptionally low density and in a unipolar photospheric magnetic field having "open" magnetic field topology. Coronal holes are largest and most stable at or near the solar poles, and are a source of high-speed (700-800 km/s) solar wind. Coronal holes are visible in several wavelengths, most notably solar x-rays visible only from space, but also in the He 1083 nm line which is detectable from the surface of the Earth. In soft x-ray images (photon energy of ~0.1 - 1.0 keV or a wavelength of 10-100 Å), these regions are dark, thus the name "holes".

Coronal Mass Ejection (CME): A transient outflow of plasma from or through the solar corona. CMEs are often but not always associated with erupting prominences, disappearing solar filaments, and flares.

Corotation (with the Earth): A plasma in the magnetosphere of the Earth is said to be corotating with the Earth if the magnetic field drags the plasma with it and together they have a 24 hour rotation period.

Cosmic Ray (galactic, solar): Extremely energetic (relativistic) charged particles or electromagnetic radiation, primarily originating outside of the Earth's magnetosphere. Cosmic rays usually interact with the atoms and molecules of the atmosphere before reaching the surface of the Earth. The nuclear interactions lead to formation of daughter products, and they in turn to granddaughter products, etc.; thus there is a chain of reactions and a "cosmic ray shower." Some cosmic rays come from outside the solar system while others are emitted from the Sun in solar flares. See also Anomalous Cosmic Ray; Energetic Particle; Solar Energetic Particle (SEP) Event.

Cross Magnetic Field Diffusion: The gradual motion (diffusion) of charged particles across magnetic fields. This is caused by collisions with other particles or by wave-particle interactions.

Cusp: In the magnetosphere, two regions near magnetic local noon and approximately 15 degrees of latitude equatorward of the north and the south magnetic poles. The cusps mark the division between geomagnetic field lines on the sunward side (which are approximately dipolar but somewhat compressed by the solar wind) and the field lines in the polar cap that are swept back into the magnetotail by the solar wind. The term cusp implies the conical symmetry around the central axis of the polar magnetic field lines. In practice, the terms "cusp" and "cleft" are often used interchangeably. However, "cleft" implies greater extension in longitude (local time) and hence a wedge-shaped structure.

Cycloidal Motion: A charged particle moves in a cycloid when the magnetic field line that it is gyrating about is convected at constant velocity.

Cyclotron Frequency: When a particle of charge q moves in a magnetic field **B**, the particle orbits, or gyrates around the magnetic field lines. The cyclotron frequency is the frequency of this gyration, and is given by $\omega_c = q|\mathbf{B}|/mc$, where m is the mass of the particle, and c is the velocity of light (in centimeter-gram-second (c.g.s.) units).

Cyclotron Resonance: The frequency at which a charged particle experiences a Doppler shifted wave at the particle's cyclotron frequency. Because the particle and wave may be traveling at different speeds and in different directions, there is usually a Doppler shift involved.

D Region: A daytime region of the Earth's ionosphere beginning at approximately 40 km, extending to 90 km altitude. Radio wave absorption in this region can be significantly increased due to increasing ionization associated with the precipitation of solar energetic particles through the magnetosphere and into the ionosphere.

Dayside Cusp: See Cusp.

Diffusion: The slow, stochastic motion of particles.

Diffusive Shock Acceleration: Charged particle acceleration at a collisionless shock due to stochastic scattering processes caused by waves and plasma turbulence. See also Shock Wave (collisionless).

Dipole Magnetic Field: A magnetic field whose intensity decreases as the cube of the distance from the source. A bar magnet's field and the magnetic field originating in the Earth's core are both approximately dipole magnetic fields.

Drift (of ions/electrons): As particles gyrate around magnetic field lines, their orbits may "drift" perpendicular to the local direction of the magnetic field. This occurs if there is a force also perpendicular to the field - e. g. an electric field, curvature in the magnetic field direction, or gravity.

Driver Gas: A mass of plasma and entrained magnetic field that is ejected from the Sun, that has a velocity higher than the upstream plasma, and which "drives" a (usually collisionless) shock wave ahead of itself.

Dst Index: A measure of variation in the geomagnetic field due to the equatorial ring current. It is computed from the H-components at approximately four near-equatorial stations at hourly intervals. At a given time, the Dst index is the average of variation over all longitudes; the reference level is set so that Dst is statistically zero on internationally designated quiet days. An index of -50 nT (nanoTesla) or less indicates a storm-level disturbance, and an index of -200 nT or less is associated with middle- latitude auroras. Dst is determined by the World Data Center C2 for Geomagnetism, Kyoto University, Kyoto, Japan.

Dust Tail: When a comet approaches the Sun, the heating of the "dirty snowball" that is the nucleus of the comet leads to the sublimation of the ice and the release of dust particles that were trapped in the ice. The trail of dust left behind is the tail. See also Ion Tail.

Dynamo (solar, magnetospheric): The conversion of mechanical energy (rotation in the case of the Sun) into electrical currents This is the process by which magnetic fields are amplified by the induction of plasmas being forced to move perpendicular to the magnetic field lines. See also Mean Field Electro-Dynamics.

E-Region: A daytime region of the Earth's ionosphere roughly between the altitudes of 90 and 160 km. The E-region characteristics (electron density, height, etc.) depend on the solar zenith angle and the solar activity. The ionization in the E layer is caused mainly by x-rays in the range 0.8 to 10.4 nm coming from the Sun.

Ecliptic Plane: The plane of the Earth's orbit about the Sun. It is also the Sun's apparent annual path, or orbit, across the celestial sphere.

Electrically Charged Particle: Electrons and protons, for example, or any atom from which electrons have been removed to make it into a positively charged ion. The elemental charge of particles is 4.8×10^{-10} esu. An electron and proton have this charge. Combined (a hydrogen atom), the charge is zero. Ions have multiples of this charge, depending on the number of electrons which have been removed (or added).

Electrojet: (1) Auroral Electrojet (AE): A current that flows in the ionosphere at a height of ~100 km in the auroral zone. (2) Equatorial Electrojet: A thin electric current layer in the ionosphere over the dip equator at about 100 to 115 km altitude.

Electron Plasma Frequency/Wave: The natural frequency of oscillation of electrons in a neutral plasma (e.g., equal numbers of electrons and protons).

Electron Volt: The kinetic energy gained by an electron or proton being accelerated in a potential drop of one Volt.

Energetic Particle: Energetic particles are defined relative to the background Thermal Plasma so that any particle having a larger energy than the thermal energy is an energetic particle. See also Cosmic Ray.

Extreme Ultraviolet (EUV): A portion of the electromagnetic spectrum from approximately 10 to 100 nm.

Extremely Low Frequency (ELF): That portion of the radio frequency spectrum from 30 to 3000 Hz.

Faculae: White-light plage - bright regions of the photosphere seen in white light, seldom visible except near the solar limb. Corresponds with concentrated magnetic fields that may presage sunspot formation.

Fast Mode (wave/speed): In magnetohydrodynamics, the fastest wave speed possible. Numerically, this is equal to the square root of the sum of the squares of the Alfvén speed and plasma sound speed.

Fibril: A linear feature in the chromosphere of the Sun as viewed in the Hydrogen alpha (Hα) line of the spectrum at 6563 Å. They occur near strong sunspots and plage or in filament channels. Fibrils parallel strong magnetic fields, as if mapping the field direction.

Field Aligned Current: A current flowing along (or opposite to) the magnetic field direction.

Filament: A mass of gas suspended over the chromosphere by magnetic fields and seen as dark ribbons threaded over the solar disk. A filament on the limb of the Sun seen in emission against the dark sky is called a prominence. Filaments occur directly over magnetic-polarity inversion lines, unless they are active.

Flare: A sudden eruption of energy in the solar atmosphere lasting minutes to hours, from which radiation and energetic charged particles are emitted. Flares are classified on the basis of area at the time of maximum brightness in H alpha.

Importance 0 (Subflare): < = 2.0 hemispheric square degrees
Importance 1: 2.1-5.1 square degrees
Importance 2: 5.2-12.4 square degrees
Importance 3: 12.5-24.7 square degrees
Importance 4: > = 24.8 square degrees

[One square degree is equal to $(1.214 \times 10E+4$ km)squared = 48.5 millionths of the visible solar hemisphere.] A brightness qualifier F, N, or B is generally appended to the importance character to indicate faint, normal, or brilliant (for example, 2B).

Flux Rope: A magnetic phenomenon which has a force-free field configuration.

Force Free Field: A magnetic field which exerts no force on the surrounding plasma. This can either be a field with no flowing electrical currents or a field in which the electrical currents all flow parallel to the field.

Free Energy (of a plasma): When an electron or ion distribution is anisotropic, they are said to have "free energy" from which plasma waves can be generated via instabilities. The waves scatter the particles so they become more isotropic, reducing the free energy.

Frozen-in Field: In a tenuous, collisionless plasma, the weak magnetic fields embedded in the plasma are convected with the plasma. i.e., the are "frozen in."

Galactic Cosmic Ray (GCR): See Cosmic Ray.

Gamma Ray: Electromagnetic radiation at frequencies higher than x-rays.

Gauss: The unit of magnetic flux in the centimeter-gram-second system; equal to 1×10^{-4} Webers per square meter in International System (SI) units.

Geomagnetic Storm: A worldwide disturbance of the Earth's magnetic field, distinct from regular diurnal variations. A storm is precisely defined as occurring when D_{ST} becomes less than -50 nT. (See geomagnetic activity).

> Initial Phase: Of a geomagnetic storm, that period when there may be an increase of the middle-latitude horizontal magnetic field intensity (H) (see geomagnetic elements) at the surface of the Earth. The initial phase can last for hours (up to a day), but some storms proceed directly into the main phase without showing an initial phase.

> Main Phase: Of a geomagnetic storm, that period when the horizontal magnetic field at middle latitudes decreases, owing to the effects of an increasing magnetospheric ring current. The main phase can last for hours, but typically lasts less than 1 day.

> Recovery Phase: Of a geomagnetic storm, that period when the depressed northward field component returns to normal levels. Recovery is typically complete in one to two days, but in cases (especially during the descending phase of the solar cycle) can take longer.

Geosynchronous Orbit: Term applied to any equatorial satellite with an orbital velocity equal to the rotational velocity of the Earth. The geosynchronous altitude is near 6.6 Earth radii (approximately 36,000 km above the Earth's surface). To be geostationary as well, the satellite must satisfy the additional restriction that its orbital inclination be exactly zero degrees. The net effect is that a geostationary satellite is virtually motionless with respect to an observer on the ground.

GeV: 10^9 electron Volts (Giga-electron Volt).

Granulation: Cellular structure of the photosphere visible at high spatial resolution. Individual granules, which represent the tops of small convection cells, are 200 to 2000 km in diameter and have lifetimes of 8 to 10 minutes.

Gyration (gryroscopic motion): The circular motion of a charged particle in a magnetic field.

Gyroradius: The radius of motion of a charged particle about a magnetic field line.

Helicity: The sense of rotation of a charged particle, plasma velocity, or magnetic field perturbation about the ambient field direction.

Heliopause: The boundary surface between the solar wind and the external galactic medium.

Helioseismology: The seismology of the Sun. The Sun has normal modes of oscillation which can be measured optically by monitoring velocity Doppler shifts.

Heliosheath: The boundary surface between the solar wind and the external galactic medium (the milky way).

Heliosphere: The magnetic cavity surrounding the Sun, carved out of the galaxy by the solar wind.

Heliospheric Current Sheet (HCS): This is the surface dividing the northern and southern magnetic field hemispheres in the solar wind. The magnetic field is generally one polarity in the north and the opposite in the south so just one surface divides the two polarities. However, the Sun's magnetic field changes over the 11 year solar sunspot cycle, and reverses polarity at solar maximum. The same thing happens in the magnetic field carried away from the Sun by the solar wind so the HCS only lies in the equator near solar minimum. It is called a "current sheet" because it carries an electrical current to balance the oppositely directed field on either side of the surface. It is very thin on the scale of the solar system - usually only a few proton gyroradii, or less than 100,000 km.

Heliotail: The tail of the heliosphere.

Helmet Streamer: See Streamer.

High Frequency (HF): That portion of the radio frequency spectrum between 3 and 30 MHz.

Instability: When an electron or ion distribution is sufficiently anisotropic, it becomes unstable (instability), generating plasma waves. The anisotropic distribution provides a source of free energy for the instability. A simple analog is a stick, which if stood on end is "unstable," but which if laid on its side is "stable." In this analog, gravity pulls on the stick and provides a source of free energy when the stick is stood on end.

Interplanetary Magnetic Field (IMF, Parker spiral): The magnetic field carried with the solar wind and twisted into an Archimedean spiral by the Sun's rotation.

Interplanetary Medium: The volume of space in the solar system that lies between the Sun and the planets. The solar wind flows in the interplanetary medium.

Interstellar (gas, neutral gas, ions, cosmic rays, wind, magnetic field, etc.): Literally, between the stars. In practicle terms, it is anything beyond the outer boundary of the solar wind (the "heliopause") yet within the milky way.

Interstellar medium (ISM): The volume of the galaxy (the milky way) lying between stars. See also Local Bubble.

Ion: 1. An electrically charged atom or molecule. 2. An atom or molecular fragment that has a positive electrical charge due to the loss of one or more electrons; the simplest ion is the hydrogen nucleus, a single proton.

Ion-Acoustic Wave: Longitudinal waves in a plasma similar to sound waves in a neutral gas. The amplitudes of electron and ion oscillations are not quite the same, and the resulting Coulomb repulsion provides the potential energy to drive the waves.

Ion Cyclotron Wave/Frequency: See Cyclotron Frequency.

Ion Tail: Neutral gas sublimates from a comet nucleus due to solar heating. Eventually, it becomes ionized by charge exchange with solar wind plasma or by solar ultraviolet radiation (the time for this to happen is on the order of 10. See also Dust Tail.

Ionization State: The number of electrons missing from an atom.

Ionosphere: The region of the Earth's upper atmosphere containing free (not bound to an atom or molecule) electrons and ions. This ionization is produced from the neutral atmosphere by solar ultraviolet radiation at very short wavelengths (<100 nm) and also by precipitating energetic particles.

Irradiance: Radiant energy flux density on a given surface (e. g. ergs cm^{-2} s^{-1}).

Isotropic Plasma: A plasma which is not anisotropic—whose properties are the same in all directions. See also Anisotropic Plasma; Plasma.

keV: 1000 electron Volts. See electron Volt.

Linear (nonlinear): In plasma and fluid parlance, small (large) amplitude in comparison to characteristic values of length, time, and wave speed in the local medium.

Local Bubble: The part of the LISM near the heliosphere. See also Interstellar Medium.

Local Fluff: The part of the local bubble nearest the heliosphere.

Local Interstellar Medium (LISM): See Interstellar Medium.

Loop (solar–loop prominence system): A magnetic loop is the flux tube which crosses from one polarity to another. A loop prominence bridges a magnetic inversion line across which the magnetic field changes direction. See also Magnetic Foot Point; Prominence.

Loss Cone Instability: An instability generated by a plasma anisotropy where the temperature perpendicular to the magnetic field is greater than the temperature parallel to the field. This instability gets its name because this condition exists in the Earth's magnetosphere and the "loss cone" particles are those that are lost into the upper atmosphere.

Luminosity (solar-total): see Solar Constant.

Magnetic Cloud: A region in the solar wind of about 0.25 AU or more in radial extent in which the magnetic field strength is high and the direction of one component of the magnetic field changes appreciably by means of a rotation nearly parallel to a plane. Magnetic clouds may be parts of the driver gases (coronal mass ejections) in the interplanetary medium.

Magnetic Diffusion: See Diffusion.

Magnetic Drift: Slow motion of magnetic field regions on the surface of a body of plasma, where a magnetic field line enters the surface.

Magnetic Foot Point: For a magnetic loop on the Sun, where the field line enters the photosphere. (See also Loop)

Magnetic Mirror: Charges particles moving into a region of converging magnetic flux (as at the pole of a magnet) will experience "Lorentz" forces that slow the particles and "mirror" them by eventually reversing their direction if the particles are initially moving slowly enough along the field line. See also Mirror Point.

Magnetic Reconnection: The act of interconnection between oppositely directed magnetic field lines.

Magnetic Storm: see Geomagnetic Storm.

Magnetopause: The boundary surface between the solar wind and the magnetosphere, where the pressure of the magnetic field of the object effectively equals the ram pressure of the solar wind plasma.

Magnetosheath: The region between the bow shock and the magnetopause, characterized by very turbulent plasma. This plasma has been heated (shocked) and slowed as it passed through the bow shock. For the Earth, along the Sun-Earth axis, the magnetosheath is about 3 Earth radii thick.

Magnetosonic Speed (acoustic speed): The speed of sound waves in a magnetized plasma. It is the equivalent of the sound speed in a neutral gas or non-magnetized plasma.

Magnetosphere: The magnetic cavity surrounding a magnetized planet, carved out of the passing solar wind by virtue of the planetary magnetic field, which prevents, or at least impedes, the direct entry of the solar wind plasma into the cavity.

Magnetotail: The extension of the magnetosphere in the antisunward direction as a result of interaction with the solar wind. In the inner magnetotail, the field lines maintain a roughly dipolar configuration. But at greater distances in the antisunward direction, the field lines are stretched into northern and southern lobes, separated by a plasmasheet. There is observational evidence for traces of the Earth's magnetotail as far as 1000 Earth radii downstream, in the antisolar direction.

Maxwell: The unit of magnetic flux in the centimeter-gram-second (c.g.s.) system.

Maxwellian Distribution: The minimum energy particle distribution for a given temperature. It is also the equilibrium distribution in the absence of losses due to radiation, collisions, etc.

Mean Field Electrodynamics: The calculation of the mean electromotive force generated by random motions. This approach is of greater rele-

vance for solar and terrestrial physics than that using a smooth velocity field. See also Dynamo.

Mean Free Path: The statistically most probably distance a particle travels
before undergoing a collision with another particle or interacting with
a wave.

Mesosphere: The region of the Earth's atmosphere between the upper limit
of the stratosphere (approximately 30 km altitude) and the lower limit
of the thermosphere (approximately 80 km altitude).

MeV: One million electron Volts. See also Electron Volt.

Mirror Point: The point where the charged particles reverse direction (mir-
rors). At this point, all of the particle motion is perpendicular to the
local ambient magnetic field. See also Magnetic Mirror.

Network (magnetic): (1) Chromospheric: a large-scale brightness pattern in
chromospheric (see chromosphere) and transition region spectral
lines, which is located at the borders of the photospheric (see photosphere) supergranulation and coincides with regions of local magnetic enhancement. These cellular patterns are typically 30,000 km across.
(2) Photospheric: a bright pattern that appears in spectroheliograms in
certain Fraunhofer spectrum lines. It coincides in gross outline with
the chromospheric network.

Nonthermal Distribution: A charged particle distribution that is non-
Maxwellian.

Particle Distribution: The distribution of actual particle energies around
the average energy for the total number of particles. This is usually
described with respect to the local magnetic field.

PCA: see polar cap absorption event.

Photosphere: The lowest visible layer of the solar atmosphere; corresponds
to the solar surface viewed in white light. Sunspots and faculae are
observed in the photosphere.

Pickup Ion: An ion which has entered the solar system as a neutral particle
and then become ionized either through charge exchange or pho-
toionization. It is called a pickup ion because as soon as the neutral
atom is ionized, it becomes attached to the magnetic field carried by
the solar wind and so is "picked up" by the solar wind.

Pitch Angle: In a plasma, the angle between the instantaneous velocity vector of a charged particle and the direction of the ambient magnetic
field.

Plage: On the Sun, an extended emission feature of an active region that is
seen from the time of emergence of the first magnetic flux until the
widely scattered remnant magnetic fields merge with the background. Magnetic fields are more intense in plage, and temperatures
are higher there than in surrounding, quiescent regions.

Plasma (ions, electrons): A gas that is sufficiently ionized so as to affect its dynamical behavior. A plasma is a good electrical conductor and is strongly affected by magnetic fields. See also Anisotropic Plasma; Isotropic Plasma.

Plasma Instability (ion, electron): When a plasma is sufficiently anisotropic, plasma waves grow, which in turn alter the distribution via wave-particle interactions. The plasma is "unstable."

Plasma Sheet: A region in the center of the magnetotail between the north and south lobes. The plasma sheet is characterized by hot, dense plasma and is a high beta plasma region, in contrast to the low beta lobes. The plasma sheet bounds the neutral sheet where the magnetic field direction reverses from Earthward (north lobe direction) to anti-Earthward (south lobe direction).

Plasma Wave (electrostatic/electromagnetic): A wave generated by plasma instabilities or other unstable modes of oscillation allowable in a plasma.

Polar Cap Absorption Event: An anomalous condition of the polar ionosphere whereby HF and VHF (3-300 MHz) radio waves are absorbed, and LF and VLF (3-300 kHz) radio waves are reflected at lower altitudes than normal. The cause is energetic particle precipitation into the ionosphere/atmosphere. The enhanced ionization caused by this precipitation leads to cosmic radio noise absorption and attenuation of that noise at the surface of the Earth. PCAs generally originate with major solar flares, beginning within a few hours of the event (after the flare particles have propagated to the Earth) and maximizing within a day or two after onset. As measured by a riometer (relative ionospheric opacity meter), the PCA event threshold is 2 dB of absorption at 30 MHz for daytime and 0.5 dB at night. In practice, the absorption is inferred from the proton flux at energies greater than 10 MeV, so that PCAs and proton events are simultaneous. However, the transpolar radio paths may still be disturbed for days, up to weeks, following the end of a proton event, and there is some ambiguity about the operational use of the term PCA.

Poloidal: The Earth's magnetic field is approximately a dipole aligned along the axis of rotation - hence, it is largely "poloidal." It would still be poloidal if the field were axisymmetric about some arbitrary direction, with the field lines lying parallel to the symmetry axis. In contrast, toroidal field would have the field lines going around the symmetry axis. See also Toroidal.

Population (of a plasma): See Particle Distribution.

Prograde: Motion around a rotating body which goes in the same direction as the rotation of that body. The Moon is in a prograde orbit around the Earth, although it revolves around the Earth only once every 29 days while the Earth rotates once every 24 hours. See also Retrograde.

Prominence: A term identifying cloud-like features in the solar atmosphere. The features appear as bright structures in the corona above the solar limb and as dark filaments when seen projected against the solar disk. Prominences are further classified by their shape (for example, mound prominence, coronal rain) and activity. They are most clearly and most often observed in H alpha. See also Loop.

Radiation Belt: Regions of the magnetosphere roughly 1.2 to 6 Earth radii above the equator in which charged particles are stably trapped by closed geomagnetic field lines. There are two belts. The inner belt's maximum proton density lies near 5000 km above the Earth's surface. Inner belt protons have high energy (MeV range) and originate from the decay of secondary neutrons created during collisions between cosmic rays and upper atmospheric particles. The outer belt extends on to the magnetopause on the sunward side (10 Earth radii under normal quiet conditions) and to about 6 Earth radii on the nightside. The altitude of maximum proton density is near 16,000-20,000 km. Outer belt protons are lower energy (about 200 eV to 1 MeV). The origin of the particles (before they are energized to these high energies) is a mixture of the solar wind and the ionosphere. The outer belt is also characterized by highly variable fluxes of energetic electrons. The radiation belts are often called the "Van Allen radiation belts" because they were discovered in 1958 by a research group at the University of Iowa led by Professor J. A. Van Allen. See also Trapped Particle.

Reconnection: A process by which differently directed field lines link up, allowing topological changes of the magnetic field to occur, determining patterns of plasma flow, and resulting in conversion of magnetic energy to kinetic and thermal energy of the plasma. Reconnection is invoked to explain the energization and acceleration of the plasmas/energetic particles that are observed in solar flares, magnetic substorms and storms, and elsewhere in the solar system.

Relativistic: Charged particles (ions or electrons) which have speeds comparable to the speed of light.

Retrograde: Motion around a rotating body which goes in the opposite direction to the rotation. See also Prograde.

Ring Current: In the magnetosphere, a region of current that flows near the geomagnetic equator in the outer belt of the two Van Allen radiation belts. The current is produced by the gradient and curvature drift of the trapped charged particles of energies of 10 to 300 keV. The ring current is greatly augmented during magnetic storms because of the hot plasma injected from the magnetotail and upwelling oxygen ions from the ionosphere. Further acceleration processes bring these ions and electrons up to ring current energies. The ring current (which is a diamagnetic current) causes a worldwide depression of the horizontal geomagnetic field during a magnetic storm.

Scintillation: Describing a degraded condition of radio propagation characterized by a rapid variation in wave amplitude and/or phase (usually on a satellite communication link) caused by variations in electron density anywhere along the signal path. It is positively correlated with ionsopheric spread F and to a lesser degree, sporadic E. Scintillation effects are the most severe at low latitudes, but can also be a problem at high latitudes, especially in the auroral oval and over the polar caps.

SEP (solar energetic particle): A Cosmic Ray of solar flare origin.

Sferics: Short for "atmospherics." Electromagnetic radiation associated with and generated in the troposphere by lightning strokes.

Sheath: See Heliosheath and Magnetosheath.

Shock Wave (collisionless): A shock wave is characterized by a discontinuous change in pressure, density, temperature, and particle streaming velocity, propagating through a compressible fluid or plasma. Collisionless shock waves occur in the solar wind when fast solar wind overtakes slow solar wind with the difference in speeds being greater than the magnetosonic speed. Collisionless shock thicknesses are determined by the proton and electron gyroradii rather than the collision lengths. See also Diffusive Shock Acceleration; Solar Wind Shock.

Solar Activity: Transient perturbations of the solar atmosphere as measured by enhanced x-ray emission (see x-ray flare class), typically associated with flares. Five standard terms are used to describe the activity observed or expected within a 24-h period:

Very low - x-ray events less than C-class.

Low - C-class x-ray events.

Moderate - isolated (one to 4) M-class x-ray events.

High - several (5 or more) M-class x-ray events, or isolated (one to 4) M5 or greater x-ray events.

Very high - several (5 or more) M5 or greater x-ray events.

Solar Constant: The total radiant energy received vertically from the Sun, per unit area per unit of time, at a position just outside the Earth's atmosphere when the Earth is at its average distance from the Sun. Radiation at all wavelengths from all parts of the solar disk are included. Its value is approximately 2.00 cal/sq cm/min = 1.37 kW/sq m and it varies slightly (by approximately 0.1%) from day to day in response to overall solar features.

Solar Corona: See Corona.

Solar Cycle: The approximately 11 year quasi-periodic variation in the sunspot number. The polarity pattern of the magnetic field reverses with each cycle. Other solar phenomena, such as the 10.7-cm solar radio emission, exhibit similar cyclical behavior. The solar magnetic field reverses each sunspot cycle so there is a corresponding 22 year solar magnetic cycle.

Solar Energetic Particle: A Cosmic Ray of solar origin.

Solar Energetic Particle (SEP) Event: A high flux event of solar cosmic rays. This is commonly generated by larger solar flares, and lasts, typically, from minutes to days. See also Cosmic Ray.

Solar Maximum: The month(s) during the sunspot cycle when the smoothed sunspot number reaches a maximum. Recent solar maxima occurred in December 1979 and July 1989.

Solar Minimum: The month(s) during the sunspot cycle when the smoothed sunspot number reaches a minimum. A recent solar minimum occurred in September 1986.

Solar Wind: The outward flow of solar particles and magnetic fields from the Sun. Typically at 1 AU, solar wind velocities are 300-800 km/s and proton and electron densities of 3-7 per cubic centimeter (roughly inversely correlated with velocity). The total intensity of the interplanetary magnetic field is nominally 3-8 nT.

Solar Wind Shock: A shock wave propagating in the solar wind. See Shock Wave.

Spiral Field: See Interplanetary Magnetic Field.

Sprite (red): Red sprites are brief optical flashes of light that emanate from the upper atmosphere coincident with, and directly above, extremely large, positive cloud-to-ground lightning strokes. The vertical extent of red sprites spans the altitude range of about 50-90 km within the mesosphere, and their horizontal dimensions are typically tens of km. The duration is only a few milliseconds. Measurements of the optical spectrum of sprites have determined that their red color is excited states of molecular nitrogen. Under the right nighttime viewing conditions red sprites can be seen with the unaided eye as brief flickers in the sky above an active thunderstorm. See also Blue Jets.

Stagnation Point: The point at which the flow speed approaches zero as a stream or air impinges on, for example, a flat wall.

Stratosphere: That region of the Earth's atmosphere between the troposphere and the mesosphere. It begins at an altitude of temperature minimum at approximately 13 km and defines a layer of increasing temperature up to about 30 km.

Streamer: A feature of the white light solar corona (seen in eclipse or with a coronagraph) that looks like a ray extending away from the Sun out to about 1 solar radius, having an arch-like base containing a cavity usually occupied by a prominence.

Subsonic: Movement at less than the local speed of sound.

Substorm: A substorm corresponds to an injection of charged particles from the magnetotail into the nightside magnetosphere. Plasma instabilities lead to the precipitation of the particles into the auroral zone ionosphere, producing intense aurorae. Enhanced ionospheric conductivity and externally imposed electric fields lead to the intensification of the auroral electrojets.

Sunspot: An area seen as a dark spot, in contrast with its surroundings, on the photosphere of the Sun. Sunspots are concentrations of magnetic flux, typically occurring in bipolar clusters or groups. They appear dark because they are cooler than the surrounding photosphere. Larger and darker sunspots sometimes are surrounded (completely or partially) by penumbrae. The dark centers are umbrae. The smallest, immature spots are sometimes called pores.

Sunspot Minimum/Maximum: See Solar Minimum/Maximum.

Supergranulation: A system of large-scale velocity cells that does not vary significantly over the quiet solar surface or with phase of the solar cycle. The cells are presumably convective in origin with weak upward motions in the center, downward motions at the borders, and horizontal motions of typically 0.3 to 0.4 km/s. Magnetic flux is more intense along the borders of the cells.

Supersonic: Above the sound speed.

Tail: See magnetotail.

Tesla: A unit of magnetic flux density (Weber/m^2).

Termination Shock: The shock wave in the solar wind which is caused by the abrupt deceleration of the solar wind as it runs into the local interstellar medium (LISM). It is thought to lie somewhere between 70 and 150 AU from the Sun.

Thermal Conductivity: In the presence of a temperature gradient, heat will flow down the gradient. The thermal conductivity measures the efficiency a material, gas, or plasma conducts heat for a given temperature gradient.

Thermal Equilibrium: A state where a gas or plasma has obtained a stable distribution or temperature through multiple collisions.

Thermal Plasma: A gas of ions and electrons, which through Coulomb collisions, is in equilibrium.

Thermal Speed (ion, electron): The random velocity of a particle associated with its temperature.

Thermosphere: That region of the Earth's atmosphere where the neutral temperature increases with height. It begins above the mesosphere at about

80-85 km and extends upward to the exosphere.

Toroidal: The toroidal direction with respect to a symmetry axis is in the azimuthal direction See also Poloidal.

Trapped Particle: Particles gyrating about magnetic field lines (e.g., in the Earth's magnetosphere). See also Radiation Belt.

Trimpi Event: Locally enhanced ionization in the auroral zone ionosphere caused by an energetic particle precipitation event. This enhanced ionization leads to absorption of cosmic radio noise as observed from the ground.

Troposphere: The lowest layer of the Earth's atmosphere, extending from the ground to the stratosphere, approximately 13 km altitude. In the troposphere, temperature decreases with height.

Ultraviolet (UV): That part of the electromagnetic spectrum between 5 and 400 nm.

Van Allen Radiation Belt: See Radiation Belt.

Very High Frequency (VHF): That portion of the radio frequency spectrum from 3 to 30 kHz.

Very Low Frequency (VLF): That portion of the radio frequency spectrum from 3 to 30 kHz.

Viscous Interaction: An interaction between two dissimilar fluids where there is a velocity gradient between them. The high beta solar wind flowing past the Earth's magnetosphere/magnetotail is one example where such an interaction can take place.

VLISM (very local interstellar medium): See Local Fluff.

Wave-particle Interaction: Particle velocity changes ("scattering") due to interaction with wave electric or magnetic fields. Since in a magnetized plasma, the particles gyrate rapidly about the magnetic field, this interaction is accomplished through a cyclotron resonant interaction (or Landau resonance).

Weber: A unit of magnetic flux equal to 10^4 Gauss m^2 (Gauss in gaussian units).

X-ray Burster: X-rays are electromagnetic radiation of very short wavelength (less than 1 nm) and very high energy; x-rays have shorter wavelengths than ultraviolet light but longer wavelengths than cosmic rays. "Soft" x-rays are those of energies less than 20 keV, or wavelengths longer than 0.05 nm.

APPENDIX: ACRONYMS AND INITIALISMS

ASCII—American Standard Code for Information Interchange

GOES—Geostationary Operational Environmental Satellite (Also called SMS/GOES)
GSFC—NASA Goddard Space Flight Center (Greenbelt, MD)

HAO—High Altitude Observatory of the National Center for Atmospheric Research, Boulder, CO
HEPAD—High Energy Proton and Alpha Detector (on GOES and TIROS)
HLMS—High Latitude Monitoring Station
HST—Hubble Space Telescope

IAGA—International Association of Geomagetism and Aeronomy
ICE—International Cometary Explorer (formerly ISEE-3), a joint NASA/ESA mission
IGY—International Geophysical Year
IMP—Interplanetary Monitoring Platform satellites
IMS—International Magnetospheric Study
ISEE-3 (see ICE)—International Sun Earth Explorer-3 satellite
ISTP—International Solar-Terrestrial Program (NASA, European Space Agency-ESA, Japanese Institute of Space and Astronautical Science-ISAS)

JPL—NASA Jet Propulsion Laboratory, Pasadena, CA

KPNO—Kitt Peak National Observatory of the National Optical Astronomy Observatories (NOAO), Tucson, AZ

MEPED—Medium Energy Proton and Electron Detector (on GOES and TIROS)
MSFC—NASA Marshall Space Flight Center, Huntsville, AL

NASA—National Aeronautics and Space Administration, Washington D.C.
NESDIS—National Environmental Satellite, Data, and Information Service, Washington D.C.
NESS—National Environmental Satellite Service, Washington D.C.
NGDC—National Geophysical Data Center
NGSDC—National Geophysical and Solar-Terrestrial Data Center, Boulder, CO

NOAA—National Oceanic and Atmospheric Administration of the Department of Commerce
NOAO—National Optical Astronomy Observatories
NRL—Naval Research Laboratory, Washington D.C.
NSF—National Science Foundation, Washington D.C.
NSO—National Solar Observatories (combines Sacramento Peak Observatory and the Solar Section of Kitt Peak Observatory)
NSSDC—National Space Science Data Center (Greenbelt, MD)

ScI—Science Institute (Space Telescope)
SEL—Space Environment Laboratory, Boulder, CO
SELDADS- Space Environment Laboratory Data Acquisition and Display System
SELSIS—Space Environment Laboratory Solar Imaging System
SEM—Space Environment Monitor (on GOES and TIROS)
SESC—Space Environment Services Center
SMM—Solar Maximum Mission
SMS—Synchronous Meteorological Satellite
SOON—Solar Observing Optical Network (USAF)
SPAN—Space Physics Analysis Network
SXI—Solar X-ray Imager

TDRS—Tracking and Data Relay Satellite (NASA)
TED—Total (particle) Energy Detector (on TIROS)
TIROS—Television and Infrared Radiation Observation Satellite

WDC—World Data Center
WMO—World Meteorological Organization